T0331174

Blockchain and Cryptocurrency

Blockchain and cryptocurrency have become the most revolutionary technologies of the 21st century, potentially transforming how we conduct business, manage assets, and exchange value. The emergence of these technologies has challenged traditional management systems and has presented new technology challenges for businesses and organizations.

Blockchain and Cryptocurrency: Management Systems and Technology Challenges explores the latest developments in blockchain and cryptocurrency and how they are changing the way we manage systems and technologies. The book delves into the technical aspects of blockchain, including its underlying architecture and consensus mechanisms, and examines the various use cases for blockchain technology, such as supply chain management, digital identity, and smart contracts. It also discusses the challenges of managing and implementing blockchain and cryptocurrency systems, including regulatory compliance, security, and scalability. Looking at the impact of these technologies on various industries, such as finance, healthcare, and energy, the book examines how they are changing the way businesses now operate.

A comprehensive guide for professionals in engineering management, business leadership, and technology that provides a comprehensive understanding of blockchain and cryptocurrencies' potential impact on organizations.

Big Data for Industry 4.0: Challenges and Applications

Series Editors: Sandhya Makkar, K. Martin Sagayam, and Rohail Hassan

Industry 4.0 or the fourth industrial revolution refers to interconnectivity, automation, and real-time data exchange between machines and processes. There is a tremendous growth in big data from the Internet of Things (IoT) and information services which drives the industry to develop new models and distributed tools to handle big data. Cutting-edge digital technologies are being harnessed to optimize and automate production including upstream supply-chain processes, warehouse management systems, automated guided vehicles, drones etc. The ultimate goal of industry 4.0 is to drive manufacturing or services in a progressive way to be faster, effective and efficient that can only be achieved by embedding modern day technology in machines, components and parts that will transmit real-time data to networked IT systems. These, in turn, apply advanced soft computing paradigms such as machine learning algorithms to run the process automatically without any manual operations.

The new book series will provide readers with an overview of the state-of-the-art in the field of Industry 4.0 and related research advancements. The respective books will identify and discuss new dimensions of both risk factors and success factors, along with performance metrics that can be employed in future research work. The series will also discuss a number of real-time issues, problems and applications with corresponding solutions and suggestions. Sharing new theoretical findings, tools and techniques for Industry 4.0, and covering both theoretical and application-oriented approaches. The book series will offer a valuable asset for newcomers to the field and practicing professionals alike. The focus is to collate the recent advances in the field, so that undergraduate and postgraduate students, researchers, academicians, and industry people can easily understand the implications and applications of the field.

Industry 4.0 Interoperability, Analytics, Security, and Case Studies
Edited by G. Rajesh, X. Mercilin Raajini, and Hien Dang

Big Data and Artificial Intelligence for Healthcare Applications
Edited by Ankur Saxena, Nicolas Brault, and Shazia Rashid

Machine Learning and Deep Learning Techniques in Wireless and Mobile Networking Systems
Edited by K. Suganthi, R. Karthik, G. Rajesh, and Ho Chiung Ching

Big Data for Entrepreneurship and Sustainable Development
Edited by Mohammed el Amine Abdelli, Wissem Ajili Ben Youssef, Ugur Ozgoker, and Imen Ben Slimene

Entrepreneurship and Big Data
The Digital Revolution
Edited by Meghna Chhabra, Rohail Hassan, and Amjad Shamim

Microgrids
Design, Challenges, and Prospects
Edited by Ghous Bakhsh, Biswa Ranjan Acharya, Ranjit Singh Sarban Singh, and Fatma Newagy

Handbook of Artificial Intelligence for Smart City Development
Management Systems and Technology Challenges

Edited by Sandhya Makkar, Gobinath Ravindran, Ripon Kumar Chakrabortty, Arindam Pal

Blockchain and Cryptocurrency
Management Systems and Technology Challenges
Edited by Asik Rahaman Jamader, Murugesan Selvam, and Biswaranjan Acharya

For more information on this series, please visit: https://www.routledge.com/Big-Data-for -Industry-4.0-Challenges-and-Applications/book-series/CRCBDICA

Blockchain and Cryptocurrency

Management Systems and Technology Challenges

Edited by
Asik Rahaman Jamader, Murugesan Selvam
and Biswaranjan Acharya

CRC Press
Taylor & Francis Group
Boca Raton London New York

CRC Press is an imprint of the
Taylor & Francis Group, an **informa** business

Designed cover image: © Vit-Mar / Shutterstock

First edition published 2025
by CRC Press
2385 NW Executive Center Drive, Suite 320, Boca Raton FL 33431

and by CRC Press
4 Park Square, Milton Park, Abingdon, Oxon, OX14 4RN

CRC Press is an imprint of Taylor & Francis Group, LLC

© 2025 selection and editorial matter, Asik Jamader, Murugesan Selvam and Biswaranjan Acharya; individual chapters, the contributors

Library of Congress Cataloging-in-Publication Data
Names: Jamader, Asik Rahaman, editor. | Selvam, Murugesan, editor. |
Acharya, Biswaranjan, editor.
Title: Blockchain and cryptocurrency: management systems and technology
challenges / edited by Asik Rahaman Jamader, Murugesan Selvam and Biswa Ranjan Acharya.
Description: Boca Raton, FL: CRC Press, 2024. | Series: Big data for
industry 4.0: challenges and applications | Includes bibliographical
references and index.
Identifiers: LCCN 2024022069 (print) | LCCN 2024022070 (ebook) |
ISBN 9781032588971 (hbk) | ISBN 9781032591322 (pbk) |
ISBN 9781003453109 (ebk)
Subjects: LCSH: Cryptocurrencies. | Blockchains (Databases)
Classification: LCC HG1710.3 .B54 2024 (print) | LCC HG1710.3 (ebook) |
DDC 658.150285/57--dc23/eng/20240819
LC record available at https://lccn.loc.gov/2024022069
LC ebook record available at https://lccn.loc.gov/2024022070

ISBN: 978-1-032-58897-1 (hbk)
ISBN: 978-1-032-59132-2 (pbk)
ISBN: 978-1-003-45310-9 (ebk)

DOI: 10.1201/9781003453109

Typeset in Times
by Deanta Global Publishing Services, Chennai, India

Contents

Preface

More than just economic, blockchain's potential benefits also span the political, humanitarian, social, and scientific spheres. In fact, some organizations have already begun to use blockchain's technological capabilities to solve pressing issues in the real world. Blockchain technology, for instance, can be used to implement in a decentralized cloud tasks that previously required administration by organizations with jurisdictional boundaries, to counter repressive political regimes. In the sensitive Edward Snowden situation, national governments prevented credit card processors from accepting donations. This is obviously helpful for transnational organizations with a neutral political stance, such as the Internet standards organization ICANN and DNS services.

Those certain manufacturing industries as well as categories can be liberated from slanted state regulatory schemes subject to the hierarchical power structures and influences of strongly backed special interest groups on governments, allowing for new disinter-mediated business models. However, outside of these instances where a public interest must transcend governmental power structures.

Blockchain and cryptocurrency have become the most revolutionary technologies of the 21st century, with the potential to transform the way we conduct business, manage assets, and exchange value. The emergence of these technologies has challenged traditional systems of management and has presented new technology challenges for businesses and organizations.

In this book, we explore the latest developments in blockchain and cryptocurrency and the ways in which they are changing the way we manage systems and technology. We delve into the technical aspects of blockchain, including its underlying architecture and consensus mechanisms, and examine the various use cases for blockchain technology, such as supply chain management, digital identity, and smart contracts. We also explore the challenges of managing and implementing blockchain and cryptocurrency systems, including regulatory compliance, security, and scalability. We look at the impact of these technologies on various industries, such as finance, healthcare, and energy, and examine the ways in which they are changing the way businesses operate.

This book is intended for managers, business leaders, and technology professionals who are looking to understand the potential and challenges of blockchain and cryptocurrency. Whether you are a seasoned professional or new to the field, this book will provide you with a comprehensive understanding of the technology and its potential implications for your organization. We hope this book will be a valuable resource for anyone interested in exploring the possibilities of blockchain and cryptocurrency and how they can be applied to improve business and management systems.

About the Editors

Asik Rahaman Jamader is working as an Assistant Professor at Pailan College of Management & Technology, Kolkata, India, approved by AICTE, Government of India, and affiliated to Maulana Abul Kalam Azad University of Technology, West Bengal. Previously he was associated as a Lecturer/Academic Counselor in the School of Management Studies, (SOMS), at Indira Gandhi National Open University, Kolkata, India, after completing his MBA from Bharathiar University, which is a State Government University of Tamil Nadu. He is a Corporate Advisory Board Member of the *Smart Journal of Business Management Studies* indexed by the Emerging Sources Citation Index, ESCI, – Web of Science, and Clarivate Analytics, with a 5.748 Impact factor. His research interests are sustainable business management and innovative techniques implemented in business management. He has 46 granted/published national and international patents/designs/copyrights and 32 journals/books/book chapters/international conference publications indexed by SCOPUS/SCIE/ESCI/WOS. He is a Section Editor of the *Smart Tourism* (ST) journal with ISSN: 2810-9821, Singapore. He is a certified reviewer of *Cogent Social Sciences*, CRC Press. Recently he joined as an ad hoc reviewer of the *International Journal of Business Intelligence Research*, IJBIR, IGI Global publishing indexed by WOS and Scopus and as an ad hoc reviewer of the *International Journal of Asian Business and Information Management*, IJABIM, IGI Global publishing indexed by WOS and Scopus. He is a notable Scientist/Inventor of West Bengal, India, as well as currently engaged with some very important research work under the guidance of Dr. Santanu Dasgupta, Principal and Associate Professor of Pailan College of Management & Technology, Kolkata, India.

Murugesan Selvam is the Member, Syndicate, Chair, School of Business Studies, Professor and Head, Department of Commerce and Financial Studies Bharathidasan University, and the Founder – Publisher-cum-Chief Editor of SMART *Journal of Business Management Studies*, a referred, indexed, professional and international journal since 2005. He is currently the Member of Senate and Syndicate, Chair for School of Business Studies, Coordinator of Centre for Entrepreneurship and Skill, Development, Coordinator – NITI Aayog (University Level) and Coordinator, Centre for Financial Market. He was the former Dean, Faculty of Management and Director i/c, UGC Academic Staff College, to his credit; Dr. Selvam has been a researcher in the area of financial management for over three decades. He has published more than 190 research papers in various referred journals including SCOPUS journals and 22 chapters in edited books and worked on more than eight Research Projects funded by UGC and ICSSR, New Delhi. He has also been instrumental in bringing out 20 Doctorates and 19 M.Phil scholars in the area of finance. He guided more than 160 PG projects/Summer Internship Projects during his career.

Biswaranjan Acharya (Senior Member, IEEE) received the MCA degree from IGNOU, New Delhi, India, in 2009, and MTech degree in computer science and engineering from the Biju Pattanaik University of Technology (BPUT), Rourkela, Odisha, India, in 2012. He is working as an Assistant Professor at the Department of Computer Engineering – AI and BDA. He has submitted his PhD thesis in computer application to the Veer Surendra Sai University of Technology (VSSUT), Burla, Odisha, India. He has a total of more than ten years of experience in academia at reputed universities, such as Ravenshaw University and in the software development field. He is the co-author of more than 70 research articles in internationally reputed journals and serves as a reviewer for many peer-reviewed journals. He has more than 50 patents to his credit. His research interests are in multiprocessor scheduling, data analytics, computer vision, machine learning, and the IoT. He is also associated with various educational and research societies, such as IACSIT, CSI, IAENG, and ISC.

Contributors

Biswaranjan Acharya
Marwadi University
India

Sarla Achuthan
India

D. Archana
Department of Computer Science
    Engineering
Nalla Malla Reddy Engineering College
India

Akash Bag
School of Law
Adamas University
India

Debesh Bhowmik
Department of Business & Accountancy
Lincoln University College
Malaysia

Biswarup Chatterjee
Pailan College of Management and
    Technology
India

Sumanta Choudhury
Pailan College of Management and
    Technology
India

S. Subrata Chowdhury
Department of Computer Science
    Engineering
SITAMS
India

Santanu Dasgupta
Pailan College of Management &
    Technology
India

S. Dhanasekar
Department of Electronics &
    Communication Engineering
Sri Eshwar College of Engineering
India

Savita Gandhi
FCAIT
GLS University
India

Kyvalya Garikapati
School of Law
Kalinga Institute of Industrial
    Technology
India

P. Geetha
School of Engineering Mohan Babu
    University
India

V. Govindaraj
Department of Electronics &
    Communication Engineering
Dr N.G.P. Institute of Technology
India

Arpit A. Jain
FCAIT
GLS University
India

Asik Rahaman Jamader
Pailan College of Management &
    Technology
India

Manickavasagam
Alagappa University
India

Baisakhi Mukherjee
Pailan College of Management and
    Technology
India

Uddipan Nath
India

Sambhabi Patnaik
School of Law
Kalinga Institute of Industrial
    Technology
India

M. Priyadharshini
Department of Computer Science
    Engineering
Nalla Malla Reddy Engineering College
India

Oum Kumari R
Jaipuria Institute of Management
India

Bharathi Ravi
College of Management
India

Alvaro Rocha
ISEG
University of Lisbon
Portugal

K. Martin Sagayam
Department of Electronics &
    Communication Engineering
Karunya Institute of Technology and
    Sciences
India

Ch. Sathyanarayana
Department of Computer Science
    Engineering
Nalla Malla Reddy Engineering College
India

Minal Shukla
Mgl Group
India

C. Venkataramanan
Department of Electronics &
    Communication Engineering
Sri Eshwar College of Engineering
India

V. Vijaya
Amity Global Business School
India

1 Blockchain Adoption for Sustainable Management System

Sumanta Choudhury and Uddipan Nath

1.1 THE FUNDAMENTALS OF BLOCKCHAIN

Blockchain functions as a real-time distributed ledger system (Panda et al., 2021). It can be visualized as a digital equivalent of a bank ledger, meticulously recording financial transactions between accounts. Unlike conventional data models in which a single entity wields control, a distributed ledger empowers multiple parties to collectively manage data following a mutually agreed-upon framework. Data is inscribed only when a consensus among the involved parties is reached, obviating the need for pre- and post-processing reconciliation. This mechanism guarantees that all stakeholders perceive identical information, promoting unanimity on the recorded facts.

Transactions are grouped into blocks, and each block accommodates a multitude of transactions (Wu et al., 2023). These blocks are intricately linked, forming an unbroken chain. In the realm of blockchain-based networks, a vast network of computers known as nodes preserves a comprehensive copy of the blockchain ledger. Remarkably, this ledger is not confined to a central authority, but instead spreads across myriad participants worldwide. Each node autonomously scrutinizes transactions adhering to predefined criteria. Only when the entire network harmonizes on a consensus does a transaction gain confirmation and find its place on the blockchain.

Blockchain gets its name from its method of storing transaction data: in blocks that link together to create a chain. As transactions increase in number, so does the blockchain itself. These blocks serve to record and verify the timing and order of transactions, which are then permanently added to the blockchain. All of this occurs within a well-defined network, governed by rules agreed upon by its participants.

Each block comprises a hash, which acts as a unique identifier (Wu et al., 2023) or digital fingerprint, along with batches of recent valid transactions marked with timestamps. Additionally, it includes the hash of the preceding block. This connection via the previous block's hash ensures that no block can be altered or inserted between two existing blocks. Consequently, each new block reinforces the validation of the previous one, fortifying the entire blockchain. This approach imbues the blockchain with tamper-evident properties, a cornerstone of its immutability.

DOI: 10.1201/9781003453109-1

1.2 BLOCKCHAIN'S IMMUNITY TO CHANGE: THE FIVE PILLARS OF IMMUTABILITY

The term "immutable" refers to something that cannot be changed, modified, or altered. When this term is applied to a blockchain, it means that the data or information stored within the blockchain is considered permanent and unchangeable once it has been recorded and confirmed in a block. Immutability is a fundamental characteristic of blockchain technology, as it ensures the integrity and trustworthiness of the data stored on the blockchain, making it resistant to unauthorized tampering or modifications. Blockchain is considered immutable primarily due to its structure and the consensus mechanism it employs. Here are the key reasons why blockchain is considered immutable (Panda et al., 2021).

- **Cryptographic Hashing**: Each block in a blockchain contains a cryptographic hash of the previous block's data, including its own hash. This forms a chain of blocks, with each block intrinsically linked to the one before it. Changing any information in a block would require recalculating the hash for that block and all subsequent blocks. This is computationally infeasible, making it extremely difficult to alter any information in a completed block without detection (Panda et al., 2021).
- **Decentralization**: Blockchains are decentralized networks, which means that multiple copies of the ledger exist across numerous nodes (computers) within the network. This decentralization ensures that no single entity or central authority has control over the entire blockchain. To alter a block, an attacker would need to compromise the majority of nodes in the network simultaneously, which is highly improbable in a well-established and secure blockchain.
- **Consensus Mechanism**: Blockchain networks rely on consensus mechanisms to validate and agree on the contents of new blocks. In the case of popular blockchains like Bitcoin and Ethereum, the consensus mechanism is proof of work (PoW), where miners compete to solve complex mathematical puzzles. Once a block is added to the blockchain, it requires the consensus of the majority of participants to alter or remove it. This consensus process adds another layer of security against tampering.
- **Transparency**: Blockchain transactions are transparent and visible to all participants in the network. Any unauthorized changes made to a block's data would be immediately noticeable to the network participants, making it nearly impossible to tamper with the information without being detected.
- **Timestamps**: Each block contains a timestamp that records the exact time when the block was added to the blockchain. This chronological order of blocks further reinforces the immutability of the data, as it becomes evident when someone tries to alter historical transactions.

These combined features create a robust and tamper-resistant system, making it exceedingly challenging to change data once it has been recorded on a blockchain.

While it is theoretically possible to alter blockchain data, doing so would require an enormous amount of computational power, control over a significant portion of the network, and the ability to evade detection, making it highly impractical and unlikely to alter blockchain data in well-established blockchain networks.

1.3 CRYPTOGRAPHY: THE BEDROCK OF BLOCKCHAIN SECURITY

Blockchains rely on an array of cryptographic principles. They encompass the safeguarding of wallets and transaction security, the fortification of consensus protocols, and the encryption of private information for anonymous accounts. Cryptography plays an indispensable role in shielding the blockchain consensus mechanism, preserving blockchain data, and ensuring the security of user accounts. It is indeed the bedrock of blockchain security.

Let's delve deeper into the critical role of cryptography in blockchain technology:

- **Security of Wallets**: Cryptography is essential for creating secure digital wallets that users employ to store their cryptocurrencies. Each wallet is associated with a pair of cryptographic keys: a public key (akin to an account number) and a private key (similar to a password). These keys use cryptographic algorithms to ensure that only the rightful owner can access and control their digital assets.
- **Transaction Security**: When a transaction is initiated on a blockchain, it undergoes cryptographic processes to ensure its integrity and security. Cryptography is used to sign transactions with the sender's private key, providing proof of ownership and preventing unauthorized alterations during transmission (Nascimento et al., 2019).
- **Consensus Protocols**: Blockchain networks rely on consensus mechanisms (Nascimento et al., 2019) to validate and add new transactions to the ledger. Cryptographic algorithms are at the core of these protocols, ensuring that network participants agree on the validity of transactions. In proof-of-work (PoW) consensus, miners solve cryptographic puzzles to add new blocks, while in proof of stake (PoS), validators are chosen based on their stake in the network.
- **Data Protection**: Blockchain records data in a secure and tamper-evident manner using cryptographic hashing (Nascimento et al., 2019). Each block contains a unique cryptographic hash of the previous block's data, creating an unbroken chain. This makes it exceedingly challenging to alter any part of the blockchain's history.
- **User Account Security**: As mentioned earlier, cryptographic keys are the foundation of user accounts in blockchain. The private key, which should be kept secret, is used to authorize transactions and access the associated assets. Public keys are used to verify the authenticity of transactions on the blockchain (Nascimento et al., 2019).
- **Anonymous Accounts**: Some blockchain platforms, like privacy coins or platforms with anonymity features (Nascimento et al., 2019), use advanced

cryptographic techniques to ensure the privacy of users. These techniques, such as zero-knowledge proofs or ring signatures, enable transactions without revealing the identity of the sender or recipient.

Blockchain technology extensively employs established cryptographic principles, including hashing, asymmetric key cryptography, and digital signatures in conjunction with fundamental principles of record-keeping. In summary, cryptography serves as the cornerstone of blockchain security by safeguarding wallets, ensuring transaction security, underpinning consensus protocols, protecting data integrity, fortifying user accounts, and enabling privacy features. Its pervasive presence within blockchain technology is vital in maintaining trust, integrity, and the immutability of distributed ledger systems. Cryptography's role extends beyond security; it also embodies the fundamental principles of decentralization and trust within the blockchain ecosystem.

1.4   INNOVATIONS AND ADVANTAGES: THE BLOCKCHAIN REVOLUTION

Blockchain technology has ushered in a new era of security, transparency, and decentralization. Its distinct advantages, including enhanced security, transparency, and reduced reliance on intermediaries, make it a transformative force across various industries. Let's understand its distinct advantages a little more elaborately.

Blockchain facilitates peer-to-peer digital data exchange without relying on third parties or intermediaries, even among untrusting parties. This data can represent various assets, such as money, contracts, land titles, medical records, or goods and services. Its potential to drive transformative changes across sectors and in the economy, industry, and society is currently under exploration by numerous organizations.

Blockchain offers robust security through cryptography, ensuring data integrity in finance and healthcare, as seen in cryptocurrencies like Bitcoin. It provides transparency, aiding supply chain management and product authenticity verification. Decentralization reduces hacking vulnerability, crucial in finance. Automation, like smart contracts, eliminates intermediaries, enhancing efficiency. Immutability and transparency build trust, as seen in voting systems. It ensures data integrity, vital in healthcare and legal sectors. Blockchain's global accessibility and innovative potential extend to various industries beyond finance, promising transformative effects.

Let's explore the distinct advantages of blockchain technology:

1. **Security**: Blockchain technology is renowned for its security features. The use of cryptography to secure data ensures that once a transaction is added to the blockchain, it becomes nearly impossible to alter or delete. This is a significant advantage, especially in industries where data integrity is paramount, such as finance and healthcare.

Example: In the world of finance, cryptocurrencies like Bitcoin use blockchain to secure digital assets. The cryptographic security ensures that Bitcoin transactions are tamper-proof, protecting users from fraud and hacking.

2. **Transparency**: Another distinct advantage of blockchain is its transparency. Every transaction added to the blockchain is visible to all participants in the network. While the identities of participants are concealed behind cryptographic addresses, the transaction history is open for anyone to inspect.

Example: In supply chain management, companies can use blockchain to track the journey of products from manufacturing to delivery. This transparency allows consumers to verify the authenticity of products and ensures ethical sourcing.

3. **Decentralization**: Decentralization is a core principle of blockchain technology. In a decentralized network, there is no single point of control or failure. This means that blockchain systems are highly resistant to censorship and hacking attempts.

Example: Traditional banking systems rely on centralized servers, making them vulnerable to cyberattacks. In contrast, blockchain-based financial systems distribute data across multiple nodes, making it much more secure.

4. **Reduced Intermediaries**: One of the revolutionary aspects of blockchain is its potential to eliminate intermediaries in various industries. Intermediaries, such as banks, payment processors, and notaries, are often necessary to facilitate and verify transactions. Blockchain can automate many of these processes, reducing costs and increasing efficiency.

Example: Smart contracts, which are self-executing contracts with the terms directly written into code, can automate tasks that would typically require intermediaries. For instance, a smart contract can automatically release payment to a seller once a product is delivered, eliminating the need for a payment processor.

5. **Trustworthiness**: Blockchain's immutability and transparency make it highly trustworthy. Since all participants have access to the same historical data, there is no need to trust a central authority. Transactions are validated through a consensus mechanism, ensuring that only valid transactions are added to the blockchain.

Example: Blockchain can be used in voting systems to enhance trust in elections. By recording votes on a blockchain, it becomes nearly impossible to tamper with the results, ensuring the integrity of the democratic process.

6. **Data Integrity**: Blockchain ensures data integrity by providing a tamper-proof record of transactions. This is especially valuable in industries where data accuracy is critical, such as healthcare and the legal sector.

Example: Electronic health records can be stored on a blockchain, ensuring that patient data remains accurate and secure. Medical professionals can access a patient's complete medical history with confidence in its integrity.

7. **Global Accessibility**: Blockchain technology operates on a global scale. Since it's not controlled by any single entity or government, it can be accessed and used by anyone with an internet connection, regardless of geographic location.

Example: Cryptocurrencies have gained popularity in countries with unstable economies, allowing people to store and transfer value without relying on their local currency.

8. **Innovation and Potential**: Blockchain's versatility and innovation potential are limitless. As developers and businesses explore its capabilities, new use cases and applications continue to emerge. This makes blockchain an exciting and ever-evolving technology.

Example: Blockchain is being explored for applications beyond finance, including art authentication, music royalties, and carbon credits trading. These innovations have the potential to transform entire industries.

1.5   CHARTING THE PATH FORWARD: NAVIGATING CHALLENGES AND OPPORTUNITIES IN BLOCKCHAIN FOR ECONOMIC DEVELOPMENT

There is a growing interest in the potential of blockchain technology to tackle enduring issues within the realm of economic development. Advocates of blockchain assert that it has the capacity to broaden avenues for exchange and cooperation, primarily by diminishing dependence on intermediaries and mitigating the associated inefficiencies. Nevertheless, it is evident that there exists a set of challenges that demand the attention of innovators in order for the technology to fully realize its potential.

While blockchain technology holds great promise for addressing economic development challenges, it also faces several challenges and obstacles that must be addressed for it to reach its full potential. Listed are some key challenges (Panda et al., 2021) that innovators must consider:

- **Scalability**: One of the foremost challenges is scalability. Public blockchains like Bitcoin and Ethereum have experienced congestion and slower transaction processing times during periods of high demand. Scaling blockchain networks to accommodate a large number of users and transactions without compromising on efficiency is a significant challenge.

- **Energy Consumption**: Many blockchain networks, especially those using proof-of-work (PoW) consensus mechanisms, consume substantial amounts of energy. This environmental impact has raised concerns, and there is a growing need for more energy-efficient consensus mechanisms.
- **Interoperability**: Different blockchain networks often operate independently, making it difficult for them to communicate and share data. Achieving interoperability between various blockchains is crucial for the seamless exchange of information and value.
- **Regulatory Uncertainty**: Blockchain technology operates in a rapidly evolving regulatory landscape. Innovators must navigate complex and sometimes conflicting regulations, which can impact the adoption and development of blockchain solutions.
- **User-Friendliness**: Blockchain and cryptocurrency interfaces can be complex for the average user. Improving user interfaces and overall user-friendliness is essential for widespread adoption, particularly in regions with limited tech literacy.
- **Privacy and Security**: While blockchain offers transparency and security, achieving the right balance between transparency and privacy is challenging. Innovators need to develop solutions that protect user data while ensuring the integrity of the blockchain.
- **Lack of Standardization**: The absence of universal standards for blockchain technology makes it difficult to develop interoperable and universally accepted solutions. Standardization efforts are ongoing, but they remain a challenge.
- **Legal and Regulatory Compliance**: Ensuring compliance with evolving legal and regulatory requirements, such as data protection and anti-money laundering (AML) regulations, is a complex task for blockchain projects.
- **Cost**: Building and maintaining blockchain networks can be expensive, and the cost of running nodes and mining operations can be prohibitive for some applications and regions.
- **Education and Awareness**: Blockchain is still a relatively new technology, and many potential users and stakeholders lack a clear understanding of its capabilities and limitations. Raising awareness and providing education about blockchain are crucial.
- **Resistance to Change**: Traditional systems and industries often resist disruption from blockchain technology. Convincing established institutions and stakeholders to embrace blockchain can be a significant challenge.
- **Environmental Concerns**: As mentioned earlier, some consensus mechanisms, like PoW, have been criticized for their energy consumption. Blockchain innovators need to find sustainable and eco-friendly alternatives.

Innovators must work collaboratively to realize blockchain's full potential in addressing economic development challenges. Progress is being made through ongoing research and development, technological advancements, regulatory frameworks, and industry collaboration. Ultimately, blockchain's success will depend on its ability to adapt and overcome these obstacles while providing tangible benefits to various sectors of the economy.

1.6 OUTLINE OF THE TOPIC

While challenges exist, blockchain's potential for innovation and positive impact on society is immense. As we continue to explore the possibilities of blockchain, it's crucial to balance its advantages with thoughtful consideration of its challenges.

It appears highly probable that blockchain technology and its various applications will maintain their rapid development trajectory, irrespective of regulatory frameworks established by governments. The ideal scenario for everyone involved, including industry stakeholders, the general populace, and governmental bodies, would involve legislators successfully finding the equilibrium between fostering innovation and ensuring public safety.

With responsible development and adoption, blockchain has the potential to revolutionize the way we transact, store data, and trust one another in the digital age.

1.7 SUSTAINABLE MANAGEMENT SYSTEMS

1.7.1 The Need for Sustainable Management: Addressing Global and Local Challenges

The world today is faced with an array of pressing challenges that threaten the well-being of both current and future generations. These challenges range from environmental crises like climate change and resource depletion to deep-rooted social inequalities. To effectively navigate this complex landscape, there arises an urgent need for sustainable management systems. The authors elucidate the gravity of global sustainability challenges and emphasize the pivotal role of sustainable management systems in addressing them.

The current global challenges of climate change, resource depletion, and social inequality necessitate urgent and comprehensive responses. Sustainable management systems provide a practical framework for addressing these challenges. They foster environmental stewardship, social equity, economic viability, and technological innovation, offering a path towards a more sustainable and equitable future. As the world grapples with the multifaceted nature of these challenges, embracing sustainable management systems becomes not just a choice but a moral and practical imperative for the well-being of humanity and the sustainability of the planet.

1.7.1.1 A Call for Sustainable Management Systems to Rebalance Our Planet

An apt metric to gauge a nation's commitment to sustainability is the ecological footprint, as assessed by the Global Footprint Network. This ecological footprint serves as a yardstick for measuring how much biologically productive land and water are required by an individual, a whole population, or an activity to produce consumed resources and absorb generated waste, given the prevailing technology and resource management practices.

In simpler terms, the ecological footprint quantifies the total greenhouse gas emissions resulting from our activities, making it a consumption-based metric. When a product, whether it's a mobile phone, a beef steak, or a bouquet of flowers, is consumed within a country, such as Denmark, Denmark's ecological footprint increases accordingly, regardless of where that product was manufactured. Ecological footprint is measured in global hectares (gha) per person and encompasses factors like built-up land, fish consumption, forest product utilization, grazing, and cropland.

Due to the global nature of trade, an individual's or a country's ecological footprint includes land and sea resources from across the world. Every country possesses a biocapacity per person, which signifies the ecosystem's ability to regenerate plant matter, alongside an ecological footprint. When these metrics are aggregated for all nations, they yield the global ecological footprint per person. In 2017, this figure stood at 2.8 gha per person. This can be juxtaposed with the global biocapacity, which amounted to 1.63 gha per person in 2017. Consequently, a global biocapacity deficit of 1.2 gha exists, underscoring the current ecological imbalance.

This ecological perspective, as indicated by metrics like the ecological footprint, underscores the pressing need for sustainable management systems. The ecological footprint reveals that our global consumption patterns are outpacing the Earth's regenerative capacity. This unsustainable trajectory is exacerbated by factors such as climate change, resource depletion, and social inequality.

Sustainable management systems are essential in addressing these challenges. They provide a structured approach to mitigating the ecological deficit highlighted by the ecological footprint. By emphasizing responsible resource management, reducing environmental impact, and promoting social equity, these systems are pivotal in achieving a more sustainable balance between human activities and the planet's capacity to support them.

Incorporating sustainable management practices into various sectors, including agriculture, industry, and energy, can help curb ecological overshoot. These systems promote responsible consumption, waste reduction, and the adoption of eco-friendly technologies. Moreover, they foster collaboration among stakeholders to ensure that economic prosperity aligns with environmental preservation and social well-being.

In essence, the ecological footprint serves as a stark reminder of the urgency to implement sustainable management systems. It highlights the unsustainable path we're currently on and underscores the critical role these systems play in redefining our relationship with the planet, addressing global sustainability challenges, and securing a more sustainable and equitable future for all.

1.7.2 SUSTAINABILITY PRINCIPLES GUIDE SUSTAINABLE MANAGEMENT SYSTEMS FOR TACKLING THE INTERCONNECTED GLOBAL CHALLENGES

In an age marked by climate change, social inequalities, and economic disparities, the principles of sustainability have emerged as guiding beacons for addressing these multifaceted challenges. Sustainability is not merely a concept; it's a mindset, a set of principles that underscore the interconnectedness of environmental, social, and economic well-being. This chapter delves into the core principles of sustainability and elucidates the pivotal role of sustainable management systems in navigating our world toward a more equitable, resilient, and sustainable future.

1.7.2.1 Principles of Sustainability: A Holistic Framework

Sustainability, at its essence, revolves around the preservation and enhancement of the quality of life for the present and future generations. This overarching objective is achieved through a comprehensive framework comprising several fundamental principles:

1. **Environmental Stewardship**: The first and perhaps the most fundamental principle of sustainability is the responsible stewardship of the environment. It acknowledges that the Earth's finite resources must be used judiciously, avoiding depletion or degradation. Sustainable practices entail reducing waste, conserving resources, protecting ecosystems, and minimizing pollution.

2. **Social Equity**: Sustainability inherently encompasses social equity, emphasizing that the benefits of economic and environmental policies must be distributed equitably among all members of society. It is a call to address social disparities, empower marginalized communities, and ensure that no one is left behind in the pursuit of a sustainable future.

3. **Economic Viability**: While environmental and social considerations are paramount, sustainability recognizes the importance of economic viability. A sustainable economy seeks to balance profitability with ecological and social responsibility. It encourages the development of businesses and industries that generate wealth while respecting planetary boundaries and societal well-being.

4. **Interconnectedness**: Sustainability principles underscore the interconnectedness of environmental, social, and economic systems. Recognizing that actions in one sphere can have profound effects on the others, sustainability strives for coherence and synergy in decision-making and policy formulation.

5. **Long-Term Perspective**: Short-term gains and immediate profits often come at the cost of long-term sustainability. A core tenet of sustainability is adopting a long-term perspective, where decisions are evaluated not just for their immediate benefits but for their consequences on future generations.

6. **Adaptability and Resilience**: In a rapidly changing world, adaptability and resilience are indispensable. Sustainable practices emphasize the ability to adapt to evolving circumstances, whether they involve environmental shifts, social dynamics, or economic fluctuations.

7. **Innovation and Technological Advancement:** Sustainability embraces innovation as a catalyst for positive change. It encourages the development and adoption of innovative technologies and practices that enhance sustainability across all sectors.

1.7.2.2   The Relevance of Sustainability Principles

The relevance of sustainability principles becomes increasingly apparent as our world grapples with pressing challenges.

1. **Environmental Challenges**: The degradation of natural ecosystems, climate change, and resource depletion are urgent environmental concerns. Sustainability principles advocate for responsible resource management, reduction in the level of pollution, and preservation of biodiversity. These principles guide initiatives to mitigate the impacts of climate change through adoption of clean energy, sustainable agriculture, and forest conservation.
2. **Social Inequities**: Persistent social inequities, ranging from income disparities to unequal access to education and healthcare, underscore the need for social equity. Sustainability principles call for inclusive and equitable development, where social well-being is prioritized, marginalized communities are uplifted, and human rights are respected.
3. **Economic Disparities**: Economic inequalities, both within and among nations, pose significant challenges. Sustainability emphasizes economic viability within planetary boundaries, promoting sustainable business practices, fair wages, and ethical supply chains. It envisions economies that prioritize well-being over wealth accumulation.
4. **Global Interconnectedness**: The COVID-19 pandemic that we experienced vividly illustrates the interconnectedness of our world. Sustainability principles underscore the importance of global cooperation and solidarity in addressing shared challenges. They call for international collaboration on issues like public health, climate change, and the conservation of biodiversity.

1.7.3   The Role of Sustainable Management Systems

Sustainable management systems serve as practical vehicles for translating sustainability principles into action. These systems are characterized by a holistic approach that integrates sustainability considerations into decision-making, operations, and governance. Here's how they address environmental, social, and economic challenges:

1. **Environmental Sustainability**:
   - **Resource Efficiency**: Sustainable management systems optimize resource use by reducing waste, conserving energy and water, and minimizing environmental impacts. They often employ eco-friendly technologies and processes to minimize the ecological footprint.
   - **Eco-Friendly Practices**: These systems foster eco-friendly practices such as recycling, sustainable sourcing, and responsible waste disposal. They are instrumental in mitigating pollution and conserving natural resources.

- **Sustainable Supply Chains**: Sustainable management systems encourage businesses to adopt sustainable supply chain practices, reducing emissions and promoting ethical sourcing.

2. **Social Equity**:
   - **Inclusivity**: Sustainable management systems prioritize inclusivity by ensuring equal opportunities for all employees, regardless of gender, race, or background. They create diverse and inclusive workplaces that reflect broader societal values.
   - **Community Engagement**: These systems engage with local communities, addressing their needs and concerns. Sustainable management seeks to be a responsible neighbor, contributing positively to the well-being of local residents.
   - **Ethical Practices**: Ethical business practices, including fair wages, safe working conditions, and respect for human rights, are central to sustainable management systems. They promote social justice and labor rights.

3. **Economic Viability**:
   - **Triple Bottom Line**: Sustainable management systems embrace the triple bottom line approach, considering economic, environmental, and social performance. They focus on sustainable revenue generation while adhering to ethical and sustainable practices.
   - **Long-Term Profitability**: These systems prioritize long-term profitability over short-term gains. By embedding sustainability into their strategies, businesses can adapt to changing market dynamics and consumer preferences.
   - **Innovation**: Sustainable management encourages innovation, leading to the development of green technologies, sustainable products, and eco-friendly services that can enhance profitability while benefiting the environment.

4. **Global Interconnectedness**:
   - **Global Collaboration**: Sustainable management systems promote international collaboration and partnerships to address global challenges such as climate change and biodiversity loss. They foster cooperation among businesses, governments, and civil society organizations on a global scale.
   - **Transparency and Reporting**: Many sustainable management systems require organizations to report on their sustainability efforts transparently. This transparency enhances accountability and allows stakeholders worldwide to assess the impact of these initiatives.

In conclusion, sustainability principles provide a comprehensive framework for addressing environmental, social, and economic challenges in our rapidly evolving world. Sustainable management systems serve as practical tools to implement these principles, guiding businesses and organizations toward a future where economic prosperity, social equity, and environmental stewardship coexist harmoniously. By embracing sustainability, we pave the way for a more resilient and equitable world, where the well-being of all, present and future, is prioritized.

1.7.4 HARMONIZING SUSTAINABLE MANAGEMENT SYSTEMS WITH UNITED NATIONS' AGENDA FOR SUSTAINABLE DEVELOPMENT – THE SUSTAINABLE DEVELOPMENT GOALS

In 2015, the United Nations embraced the Sustainable Development Goals (SDGs), also known as the Global Goals, aiming for a universal endeavor to eradicate poverty, safeguard the environment, and secure a future where everyone experiences peace and prosperity by 2030 (United Nations, 2021).

These 17 SDGs function in unison, acknowledging that actions taken in one domain ripple through to others, underscoring the need to harmonize social, economic, and environmental sustainability in development efforts. Nations have pledged to prioritize advancement for those who are the farthest from progress, with the SDGs designed as a blueprint to combat poverty, hunger, AIDS, and gender discrimination, particularly against women and girls (United Nations, 2021).

Sustainable management systems are intricately linked with the United Nations' prescribed SDGs and are a practical and strategic approach for businesses and organizations to contribute to the achievement of the SDGs. Here is how they are connected:

- **Alignment with SDG Objectives**: The SDGs are a set of 17 Global Goals aimed at addressing some of the world's most pressing challenges, including poverty, inequality, climate change, and environmental degradation. Sustainable management systems are designed to align an organization's activities, strategies, and operations with these SDG objectives. They provide a framework for businesses to work toward common Global Goals.
- **Incorporating Sustainability Principles**: Sustainable management systems, such as ISO 14001 (Environmental Management) and ISO 26000 (Social Responsibility), encourage organizations to integrate sustainability principles into their decision-making processes. This includes considerations related to environmental impact, social responsibility, ethical practices, and economic sustainability – all of which are key components of the SDGs.
- **Measuring Progress**: Sustainable management systems often involve the establishment of key performance indicators (KPIs) and metrics to track and measure progress in sustainability efforts. These KPIs can be aligned with specific SDGs to assess how an organization's actions contribute to achieving those goals.
- **Transparency and Reporting**: Many businesses using sustainable management systems are required or choose to report on their sustainability efforts. This includes disclosing their contributions to SDGs through

sustainability reports and other means, which promotes transparency and accountability in their sustainability initiatives.

- **Stakeholder Engagement**: Sustainable management systems emphasize stakeholder engagement and dialogue, including partnerships with various stakeholders such as governments, NGOs, and local communities. Collaboration is essential to achieving many of the SDGs, as it often requires collective action.
- **Supply Chain Impact**: Sustainable management systems often extend beyond the organization itself to its supply chain partners. Organizations are encouraged to promote sustainability principles and practices among their suppliers, thereby extending the positive impact on sustainable development beyond their immediate operations.
- **Social Responsibility:** Many of the SDGs are directly related to social issues, such as poverty reduction, quality education, gender equality, and decent work. Sustainable management systems include elements of social responsibility, promoting fair labor practices, human rights, and community engagement, all of which align with specific SDGs.

In essence, sustainable management systems provide a structured approach for organizations to integrate sustainability into their core operations, and this integration naturally contributes to the advancement of the 17 SDGs. By adopting sustainable management systems, businesses and institutions play a vital role in addressing global challenges and working toward a more sustainable and equitable future as outlined by the Global Goals – eradicating poverty, safeguarding the environment, and securing peace and prosperity for all by 2030.

1.7.5   GLOBAL CHALLENGES TEST SDGs AS 2030 NEARS – DIGITAL TECHNOLOGIES OFFER A BEACON OF HOPE

Approaching the halfway mark to the 2030 deadline, the Sustainable Development Goals (SDGs), initially ambitious in 2015, now face even greater challenge (United Nations Sustainable Development Goals Report, 2022). The COVID-19 pandemic, climate crises, and complex international conflicts have not only decelerated progress but have, in certain instances, reversed advancements across multiple SDGs. Shockingly, the year 2020 witnessed a surge in the global poverty rate for the first time in over 20 years.

Amidst growing global turbulence, 90% of countries have experienced unprecedented setbacks in human development progress. With merely seven years remaining to realize the 2030 Agenda, none of the 17 SDGs has been fully realized. In simpler terms, unless immediate action is taken the 2030 Agenda may remain an unfulfilled vision for a world on the brink.

As we stand at this critical juncture, digital technologies have emerged as potential saviors in rekindling progress toward the SDGs. From ubiquitous mobile phones to cutting-edge innovations harnessing artificial intelligence and other technologies, these digital tools could prove instrumental in revitalizing the faltering journey toward the SDGs.

In an increasingly complex world where progress against many SDGs has faltered or even reversed, these digital innovations offer a glimmer of hope. To borrow the words of UN Secretary-General Antonio Guterres, "Unless we act now, the 2030 Agenda will become an epitaph for a world that might have been."

1.7.5.1 Illustrative Cases of Blockchain Applications for Addressing the SDGs

The authors have visualized certain use cases that illustrate the diverse ways in which blockchain technology can be leveraged to address each of the SDGs, in turn promoting sustainable development and positive social impact. Here are some illustrative use cases of blockchain applications toward addressing the SDG (Table 1.1).

1.8 INTEGRATION OF BLOCKCHAIN AND SUSTAINABLE MANAGEMENT

Blockchain technologies offer a valuable toolkit for businesses (Ashurst et al., 2022) that aim to align their operations with sustainability goals. These technologies can indeed prove useful to businesses that are willing to explore ways to reduce environmental impact, promote social responsibility, and ensure financial viability while delivering products and services. Blockchain offers several features and capabilities that can support sustainability initiatives and responsible business practices, which include:

- **Transparency and Traceability**: Blockchain's transparent and immutable ledger allows businesses to trace the origins and journey of products and raw materials throughout the supply chain. This transparency helps identify inefficiencies, track environmental impact, and verify the authenticity of sustainable claims. For example, it can be used to trace the origin of sustainably sourced materials, such as ethically mined minerals or organic food products.
- **Supply Chain Sustainability:** Blockchain can enhance supply chain sustainability by enabling real-time monitoring of environmental factors. Sensors and Internet of Things (IoT) devices connected to the blockchain can track temperature, humidity, and other environmental conditions during the transportation of goods, ensuring compliance with sustainability standards. This reduces waste, spoilage, and the carbon footprint of supply chains.
- **Reducing Fraud and Counterfeits:** Blockchain's secure and tamper-proof records help in the fight against counterfeit products. Businesses can use blockchain to verify the authenticity of products and ensure that consumers are getting genuine, sustainable, and ethically sourced items. This fosters trust and social responsibility.

TABLE 1.1
Possible Use Case

SDG	Possible Use Case
1. No Poverty	Blockchain can be used to create decentralized financial systems that provide access to banking services for unbanked and underprivileged populations, enabling them to save, borrow, and build credit.
	Blockchain-based microfinance platforms can provide access to financial services for individuals in poverty, enabling them to borrow and save money securely.
2. Zero Hunger	Blockchain can help improve supply chain transparency, allowing consumers to track the origin and journey of food products, ensuring food safety, reducing waste, and eliminating fraud.
	Blockchain can be used to create a transparent and traceable food supply chain, ensuring the safety and authenticity of food products, reducing waste, and preventing food fraud.
3. Good Health and Well-being	Blockchain can enhance the sharing and security of medical records, ensuring interoperability between healthcare providers, and enabling patients to have control over their health data.
	Blockchain can be used to create a transparent and traceable food supply chain, ensuring the safety and authenticity of food products, reducing waste, and preventing food fraud.
4. Quality Education	Blockchain can provide verifiable and tamper-proof certifications and credentials, ensuring the authenticity of educational achievements and improving the process of verifying academic qualifications.
	Blockchain can facilitate the issuance and verification of digital credentials and certificates, making it easier for employers and educational institutions to verify qualifications and achievements.
5. Gender Equality	Blockchain can create decentralized identity systems, empowering individuals to have control over their personal data, which can help address issues related to gender-based discrimination and identity theft.
	Blockchain can support digital identity systems that empower individuals, particularly women, to have control over their personal data and protect against identity theft and discrimination.
6. Clean Water and Sanitation	Blockchain can be utilized to monitor and manage water resources, ensuring transparency, efficient distribution, and accountable water management practices.
	Blockchain can enhance water management systems by tracking water usage, quality, and distribution, ensuring efficient and transparent management of water resources.

(Continued)

TABLE 1.1 (CONTINUED)
Possible Use Case

SDG	Possible Use Case
7. Affordable and Clean Energy	Blockchain can facilitate peer-to-peer energy trading, allowing individuals and communities to trade surplus energy generated by renewable sources, promoting the adoption of clean energy technologies.
	Blockchain can enable peer-to-peer energy trading, allowing individuals and communities to buy and sell excess renewable energy, facilitating the transition to clean and sustainable energy sources.
8. Decent Work and Economic Growth	Blockchain can enable transparent and secure digital marketplaces, reducing fraud and providing fair compensation to freelancers and gig workers.
	Blockchain-based platforms can provide fair and transparent contracts and payments for freelancers and gig workers, ensuring fair compensation and reducing exploitation.
9. Industry, Innovation, and Infrastructure	Blockchain can improve supply chain management by enhancing traceability, reducing counterfeiting, and ensuring the ethical sourcing of materials and products.
	Blockchain can improve supply chain management by enhancing transparency and traceability, reducing counterfeiting, and ensuring the ethical sourcing practices.
10. Reduced Inequalities	Blockchain can enable transparent and accountable distribution of aid and resources, ensuring that they reach the intended recipients and reducing corruption.
	Blockchain can enable the secure and transparent distribution of aid and resources, ensuring they reach the intended beneficiaries and reducing corruption and inequality.
11. Sustainable Cities and Communities	Blockchain can support smart city initiatives by facilitating secure and efficient data exchange between various urban systems, enabling better management of resources and services.
	Blockchain can support smart city initiatives by enabling secure and efficient data exchange between different urban systems, improving resource management, and enhancing citizen participation.
12. Responsible Consumption and Production	Blockchain can track and verify the entire lifecycle of products, from raw materials to disposal, promoting sustainable practices and reducing waste.
	Blockchain can track and verify the sustainability and authenticity of products, providing consumers with information about the environmental impact and ethical practices associated with their purchases.

(Continued)

TABLE 1.1 (CONTINUED)
Possible Use Case

SDG	Possible Use Case
13. Climate Action	Blockchain can facilitate the trading and tracking of carbon credits, encouraging emission reduction efforts and supporting the transition to a low-carbon economy.
	Blockchain-based carbon credit platforms can facilitate the trading and tracking of carbon credits, incentivizing emission reduction efforts and supporting climate change mitigation.
14. Life Below Water	Blockchain can aid in monitoring and protecting marine ecosystems by tracking and verifying the sustainability and legality of seafood products. Blockchain can help monitor and track fishing activities, ensuring sustainable practices and preventing illegal, unreported, and unregulated fishing.
15. Life on Land	Blockchain can help combat illegal logging and deforestation by providing transparency and traceability in the timber supply chain.
	Blockchain can facilitate the transparent and traceable tracking of timber and other natural resources, reducing illegal logging and supporting sustainable forest management.
16. Peace, Justice, and Strong Institutions	Blockchain can provide secure and immutable records of legal transactions, enhancing transparency and reducing corruption in governance and legal systems.
	Blockchain can provide secure and transparent records of legal transactions, promoting trust and integrity in governance systems and reducing corruption.
17. Partnerships for the Goals	Blockchain can enable the secure and efficient collaboration and data sharing among multiple stakeholders, fostering partnerships and cooperation towards achieving the SDGs.
	Blockchain can enable the secure and decentralized collaboration and data sharing among multiple stakeholders, fostering partnerships and cooperation towards achieving the SDGs.

- **Smart Contracts for Sustainability Agreements:** Smart contracts, which automatically execute predefined actions when certain conditions are met, can be employed to create sustainability agreements. For example, a smart contract could release payment to a supplier only when certain sustainability metrics are achieved, encouraging sustainable practices throughout the supply chain.
- **Carbon Emission Tracking**: Blockchain can be used to track and manage carbon emissions. By recording carbon credits, emissions data, and offset

transactions on a blockchain, businesses can transparently manage their carbon footprint and engage in carbon offset initiatives more effectively.

- **Fair Trade and Ethical Labor**: Blockchain can verify fair trade and ethical labor practices by recording labor conditions, wages, and certifications on the blockchain. Consumers can then verify that the products they purchase are produced in socially responsible ways.
- **Renewable Energy Trading**: Blockchain can facilitate peer-to-peer trading of renewable energy. This encourages the use of clean energy sources and allows businesses to reduce their reliance on fossil fuels, promoting environmental sustainability.
- **Environmental Impact Reporting:** Blockchain can simplify the process of collecting and reporting environmental impact data, making it easier for businesses to comply with sustainability regulations and demonstrate their commitment to stakeholders.

In summary, by leveraging blockchain's transparency, security, and automation capabilities, companies can reduce environmental impact, promote social responsibility, and ensure financial viability while delivering products and services that meet the demands of environmentally and socially conscious consumers.

1.8.1   POTENTIAL SYNERGIES BETWEEN BLOCKCHAIN TECHNOLOGY AND SUSTAINABLE MANAGEMENT

Blockchain technologies can prove highly useful to businesses that are striving to address issues like labor conditions, community engagement, resource conservation, and cost-effectiveness. Blockchain's inherent features and capabilities can support efforts to improve these areas of business operations.

- **Supply Chain Transparency**: Blockchain can provide complete transparency into the supply chain, allowing businesses to trace the source of materials and products. This transparency can help identify and address issues related to labor conditions, such as exploitation or unsafe working conditions, as well as ensure that suppliers adhere to ethical and fair labor practices.
- **Provenance Tracking**: Blockchain enables the tracking of products and materials from origin to end-users. This not only ensures product authenticity but also promotes resource conservation by reducing waste and inefficiencies in the supply chain.
- **Smart Contracts for Fair Labor Practices**: Smart contracts can be used to automate and enforce fair labor practices within the supply chain. For instance, payments to suppliers or workers can be automatically triggered upon verification of compliance with labor standards.
- **Community Engagement and Fair Trade**: Blockchain can empower local communities by providing them with more control and transparency over their resources and products. It allows for the creation of systems that

ensure fair compensation for local producers and encourages community engagement in sustainable practices.

- **Verification of Sustainable Practices**: Blockchain can be used to record and verify sustainable and eco-friendly practices, such as responsible forestry or organic farming. This can help businesses prove their commitment to resource conservation and sustainability to consumers and stakeholders.
- **Resource Management**: IoT devices and sensors connected to blockchain networks can monitor resource usage, such as water and energy consumption. This data can be used to optimize resource management, reduce waste, and lower operational costs while promoting resource conservation.
- **Community Tokens and Incentives**: Some blockchain platforms enable the creation of community tokens or tokens that represent environmental benefits. Businesses can use these tokens to incentivize and reward sustainable practices among employees, suppliers, and local communities.
- **Auditing and Compliance**: Blockchain's transparency and immutability make it an effective tool for auditing and compliance purposes. It can streamline the verification of labor conditions, environmental impact, and other compliance-related data, reducing administrative costs and errors.
- **Cost-Effective Transactions**: Blockchain's efficiency in recording and verifying transactions can result in cost savings, which can be redirected toward sustainability initiatives.

1.8.2 USE OF BLOCKCHAIN ADOPTION IN SUSTAINABLE MANAGEMENT

Blockchain technologies can also prove to be useful to businesses that are willing to explore ways to reduce environmental impact, promote social responsibility, and ensure financial viability while delivering products and services. Blockchain offers several features and capabilities that can support sustainability initiatives and responsible business practices, which include:

- **Transparency and Traceability**: Blockchain's transparent and immutable ledger allows businesses to trace the origins and journey of products and raw materials throughout the supply chain. This transparency helps identify inefficiencies, track environmental impact, and verify the authenticity of sustainable claims. For example, it can be used to trace the origin of sustainably sourced materials, such as ethically mined minerals or organic food products.
- **Supply Chain Sustainability**: Blockchain can enhance supply chain sustainability by enabling real-time monitoring of environmental factors. Sensors and IoT devices connected to the blockchain can track temperature, humidity, and other environmental conditions during the transportation of goods, ensuring compliance with sustainability standards. This reduces waste, spoilage, and the carbon footprint of supply chains.
- **Reducing Fraud and Counterfeits**: Blockchain's secure and tamper-proof records help in the fight against counterfeit products. Businesses can use

blockchain to verify the authenticity of products and ensure that consumers are getting genuine, sustainable, and ethically sourced items. This fosters trust and social responsibility.

- **Smart Contracts for Sustainability Agreements**: Smart contracts, which automatically execute predefined actions when certain conditions are met, can be employed to create sustainability agreements. For example, a smart contract could release payment to a supplier only when certain sustainability metrics are achieved, encouraging sustainable practices throughout the supply chain.
- **Carbon Emission Tracking**: Blockchain can be used to track and manage carbon emissions. By recording carbon credits, emissions data, and offset transactions on a blockchain, businesses can manage their carbon footprint in a transparent manner and engage in carbon offset initiatives more effectively.
- **Fair Trade and Ethical Labor**: Blockchain can verify fair trade and ethical labor practices by recording labor conditions, wages, and certifications on the blockchain. Consumers can then verify that the products they purchase are produced in socially responsible ways.
- **Renewable Energy Trading**: Blockchain can facilitate peer-to-peer trading of renewable energy. This encourages the use of clean energy sources and allows businesses to reduce their reliance on fossil fuels, promoting environmental sustainability.
- **Environmental Impact Reporting**: Blockchain can simplify the process of collecting and reporting environmental impact data, making it easier for businesses to comply with sustainability regulations and demonstrate their commitment to stakeholders.

Blockchain adoption also offers immense potential for the development of sustainable management systems. By leveraging the technology's transparency and immutability, supply chains can be monitored and validated, promoting ethical and sustainable practices. Here are some illustrative use cases of blockchain adoption:

- **Decentralized Energy Grids**: Blockchain can enable the creation of decentralized energy grids and enhance carbon credits and emissions trading. As the world grapples with pressing environmental challenges, the adoption of blockchain technology becomes imperative for a sustainable future.
- **Renewable Energy Management**: One area where blockchain can be useful is in renewable energy management. With the increasing demand for clean energy, it is essential to ensure that the energy produced is indeed from renewable sources. Blockchain can create a system where each unit of energy generated is recorded on the blockchain, along with its source. This will enable consumers and organizations to verify the origin of the energy they are using, promoting the adoption of renewable sources.
- **Supply Chain Management**: Another area where blockchain can bring about positive change is in supply chain management. Many products today

claim to be environmentally friendly or sustainably sourced. However, verifying these claims can be challenging. By implementing blockchain in the supply chain, every step of the production process can be recorded, ensuring transparency and traceability. This will enable consumers to make informed choices and support companies that adhere to sustainable practices.

- **Emissions Trading and Carbon Credit and Market:** By using block-chain, carbon emissions can be accurately recorded and verified, ensuring that companies meet their emissions targets. This technology also enables the creation of digital tokens that represent units of carbon emissions, which can be traded on a global marketplace. This incentivizes companies to reduce their carbon footprint and provides a transparent and efficient way to manage emissions reductions. Additionally, blockchain can facilitate the establishment of a carbon credit market. Carbon credits are tradable units that represent a reduction in greenhouse gas emissions. By using block-chain, the creation, trading, and retirement of carbon credits can be tracked efficiently, ensuring accountability and preventing fraud.

1.9 CONCLUSION

Rooted in transparency, immutability, and trustlessness, blockchain offers innovative solutions across industries and to governments. To fully harness its potential, we must prioritize tackling scalability and energy consumption challenges. Ongoing discussions and research are shaping the blockchain future.

Embracing blockchain is pivotal for advancing sustainable management and the UN's 17 SDGs. Yet, success demands addressing hurdles and embracing a holistic approach. This journey involves technology, society, and regulations. Engaging in ongoing discussions and initiatives is our path to a more sustainable and equitable world.

Key summary points include:

- Blockchain technology is disrupting traditional data management and security paradigms by offering decentralization, transparency, immutability, and trustlessness.
- Transparency is a core feature of blockchain, providing a tamper-resistant audit trail for all transactions, enhancing accountability, and reducing fraud.
- Immutability ensures that once a transaction is added to the blockchain, it cannot be altered or erased, maintaining trustworthy historical records.
- Trustlessness means blockchain participants don't rely on central authorities to validate transactions, reducing the risk of human error and bias.
- Blockchain's adoption aligns with the UN's Sustainable Development Goals (SDGs) by enhancing supply chain transparency, supporting ethical sourcing, and advancing goals related to responsible consumption, peace, justice, and strong institutions.

- Immutability in blockchain facilitates secure ownership rights and thereby fosters trust in institutions.
- Blockchain promotes financial inclusion through DeFi platforms.
- Challenges such as scalability, energy consumption, and regulatory frameworks must be addressed for blockchain's successful integration into sustainable management systems.
- Blockchain's potential is significant but not a standalone solution; it must be integrated responsibly within a broader context of governance and collaboration.

Ongoing discussions and research continue to shape the future of blockchain technology and its role in advancing sustainable and equitable global development.

REFERENCES

Anderberg, A., Andonova, E., Bellia, M., CalÃ¨s, L., Inamorato Dos Santos, A., Kounelis, I., NaiFovino, I., PetraccoGiudici, M., Papanagiotou, E., Sobolewski, M., Rossetti, F. and Spirito, L. (2019). *Blockchain Now and Tomorrow* (Figueiredo Do Nascimento, S. and Roque Mendes Polvora, A., Eds.). EUR 29813 EN, Publications Office of the European Union, Luxembourg, ISBN 978-92-76-08977-3, doi:10.2760/901029, JRC117255.

Ashurst, S. and Tempesta, S. (2022). *Blockchain Applied Practical Technology and Use Cases of Enterprise Blockchain for the Real World*. Routledge.

Asian Development Bank. "Promoting Financial Inclusion through Distributed Ledger Technology." Asian Development Bank, 08 Jan. 2018, https://development.asia/explainer/promoting-financial-inclusion-through-distributed-ledger-technology. Accessed 20 Sept. 2023.

Asian Development Bank. "Using Blockchain to Improve Aid Transparency and Efficiency." Asian Development Bank, 23 Feb. 2023, https://development.asia/case-study/using-blockchain-improve-aid-transparency-and-efficiency. Accessed 20 Sept. 2023.

Bank for International Settlements. (n.d.). "Central bank digital currencies." Bank for International Settlements. Accessed [30 September 2023]. https://www.bis.org/cpmi/publ/d174.pdf

Chishti, S. and Barberis, J. (2016). *The FinTech Book*. Wiley. ISBN 978-1-119-21887-6.

Crane, A., Palazzo, G., Spence, L., and Matten, D. (2014). Contesting the Value of 'Creating Shared Value'. *California Management Review*, 56, 130–153. doi:10.1525/cmr.2014.56.2.130.

"Distributed Ledger Technologies for Developing Asia." Asian Development Bank, Accessed [20 September 2023], https://www.adb.org/sites/default/files/publication/388861/ewp-533.pdf.

"Enter the Triple Bottom Line." John Elkington's Website. Accessed [29 September 2023]. https://www.johnelkington.com/archive/TBL-elkington-chapter.pdf.

Epstein, M., Buhovac, A., Elkington, J. and Leonard, H. (2017). Making Sustainability Work: Best Practices in Managing and Measuring Corporate Social, Environmental, and Economic Impacts. doi:10.4324/9781351276443.

European Commission. (2022). Overview of EU-funded blockchain-related projects. https://digital-strategy.ec.europa.eu/en/news/overview-eu-funded-blockchain-related-projects.

Investopedia. (n.d.). "Triple Bottom Line." Investopedia. Accessed [29 September 2023]. https://www.investopedia.com/terms/t/triple-bottom-line.asp.

Iron Mountain. "What Is Blockchain and Why Should Records Management Professionals Care?" Ironmountain.com. Accessed [19 September 2023]. https://www.ironmountain .com/resources/general-articles/w/what-is-blockchain-and-why-should-records-man-agement-professionals-care.

Kreye, M. E. (2023). *Sustainable Operations and Supply Chain Management*. Routledge.

MSCI. (2021). *The Blockchain Future*. MSCI. https://www.msci.com/documents/1296102 /28401354/ThematicIndex-Blockchain-cbr-en.pdf.

Nascimento, S. (ed.), Pólvora, A. (ed.), Anderberg, A., Andonova, E., Bellia, M., Calès, L., Inamorato dos Santos, A., Kounelis, I., Nai Fovino, I., Petracco Giudici, M., Papanagiotou, E., Sobolewski, M., Rossetti, F. and Spirito, L. (2019). *Blockchain Now and Tomorrow: Assessing Multidimensional Impacts of Distributed Ledger Technologies*, EUR 29813 EN, Publications Office of the European Union, Luxembourg. ISBN 978-92-76-08977-3, doi:10.2760/901029, JRC117255.

Panda, S. K., Jena, A. K., Swain, S. K., and Satapathy, S. C. (2021). *Blockchain Technology: Applications and Challenges*. Springer Nature Switzerland AG.

Pisa, M. and Juden, M. (2017). *Blockchain and Economic Development: Hype vs. Reality*. CGD Policy Paper. Center for Global Development. https://www.cgdev.org/publication /blockchain-and-economic-development-hype-vs-realit.

Polvora, A., Brekke, J. K., and Hakami, A. (2021). *Distributed Ledger Technologies (DTLs) for Social and Public Good – Where to next*. European Commission, Brussels, JRC127939.

Roopika J. (2020). "Blockchain Technology: History, Concepts, and Applications." *IRJET* 7, no. 10. https://www.irjet.net/archives/V7/i10/IRJET-V7I10109.pdf.

UN. (2021). Strengthening the Multi-Stakeholder Dimension of National Development Planning and SDG Mainstreaming [PDF file]. https://sdgs.un.org/sites/default/files /2021-11/Final%20Strengthening%20the%20multi-stakeholder%20dimension%20of %20national%20development%20planning%20and%20SDG%20mainstreaming.pdf.

Wu, B. and Wu, B. (2023). *Blockchain for Teens With Case Studies and Examples of Blockchain Across Various Industries*. Apress Media LLC.

World Health Organization (WHO). (n.d.). *Zoonoses*. WHO. Accessed [24 September 2023]. https://www.who.int/news-room/fact-sheets/detail/zoonoses.

2 Significance and Consequence of Cryptocurrency

M. Priyadharshini, D. Archana,
Ch. Sathyanarayana, S. Subrata Chowdhury,
and Alvaro Rocha

2.1 INTRODUCTION

There has been a paradigm shift in the way transactions are being conducted—from physical payments like cash and checks toward digital transactions in an environment where technology is constantly changing. Predicting the projected value of a currency is crucial to utilizing it as an asset or a transactional medium. The value of any currency and its stability are determined mainly by the organization in charge; for fiat currencies, this body is represented by the national government. Harmful government meddling in the financial system can have unexpected, grave consequences. However, the value of digital currencies depends on thetrustworthiness and security of the platform they are employed on. Double-spending is a vulnerability to conventional digital money. Cyberspace security flaws might result in manipulation of transaction data risk digital currencies. Traditional currencies are more susceptible to volatility and devaluation due to these weaknesses. Using cryptocurrency that is based on blockchain offers a potential answer to the problems stated above. To guarantee security, decentralization, and transparency—essential for a successful currency—blockchain, an emerging technology, maintains data immutability over a network. Contrary to traditional currency, cryptocurrency transactions are encrypted using cryptographic ciphers. Cryptocurrencies have been at the forefront of this forward-thinking innovation for the past ten years, and digital banking has risen dramatically (Mourgayar et al., 2016).

According to CAGR projections, the market capitalization of cryptocurrencies is estimated to be $266 billion, with a growth rate of 11.9% expected by 2024. A cryptocurrency's key characteristic is its decentralized nature, which it receives from the blockchain. It prevents corruption because a single entity cannot control it. Therefore, cryptocurrencies are intrinsically resistant to devaluations brought on by corruption, as opposed to those that may occur in fiat currency. In the blockchain network, cryptocurrencies prevent the problem of double-spending by receiving several

DOI: 10.1201/9781003453109-2

confirmations from adjacent nodes. The more information there is, the more secure and irrevocable the transaction is. Since changing a record across all network nodes is practically impossible, the blockchain ledger's transactional documents cannot be altered. Since the record cannot be modified after a successful transaction, it cannot be changed either (Kili Carslan et al., 2023).

Cryptocurrencies may be used as a store of value and a transactional means of exchange owing to its aforementioned positive traits and universal accessibility. Nevertheless, unstable market patterns and popular opinion continue to impact how much value cryptocurrencies retain significantly. Additionally, traditional financial procedures are useless because of the poor connection between significant financial assets and cryptocurrencies. The Pearson relationship between Bitcoin and other crucial financial support is discussed. There is no correlation between the exchange rate between Bitcoin and the cost of gold. Since many publicly available data on cryptocurrency market and societal trends exist, machine-learning approaches are utilized to estimate Bitcoin prices. These algorithms are employed for concluding mathematical models from data devoid of any explicit programming of the computer to perform certain task. When data complexity for Bitcoin market rises, the models can acquire complex data representations. Recurrent neural networks, a subset of deep learning models, can handle the time-series challenge of forecasting cryptocurrency values. Several writers have predicted the securities issues and equities with deep and machine learning algorithms in recent years (Strilets et al., 2022). However relatively very little studies has been conducted on predicting the price of cryptocurrencies.

The primary function of cryptocurrencies is as a payment method instead of using cash. Nonetheless, in recent years, it has been clear that investing in cryptocurrency trading may be quite profitable (Kaur et al., 2018). The Bitcoin market is viewed by many stock market analysts and other investors as the most dangerous sector to invest in because of its volatility and strong dependence on public opinion. We will also be able to calculate the total processing power of the blockchain by precisely predicting the value of various cryptocurrencies. With a wide range of variables, including societal attitudes and governmental policies, the value of cryptocurrencies is greatly influenced by previous price trends and transaction volumes. Many recent studies have examined how to predict cryptocurrency values. Based on the opinions stated above, we provide a discussion that may be used to forecast the cost of further coins (Murthy et al., 2020). To imitate market volatility, we induce various approaches to address the issue of unpredictable changes in cryptocurrency values.

2.2   BLOCKCHAIN FRAMEWORK

The node starts transaction by applying digital signature that uses private key cryptography over decentralized blockchain network (as seen in Figure 2.1 above). One way to conceptualize a transaction on the blockchain network is as a data format that symbolizes peer-to-peer transfers of digital goods. Once all transactions have been added, they are spread over the network via the flooding technique known as the Gossip protocol, an unconfirmed transaction pool. Then, according to preset

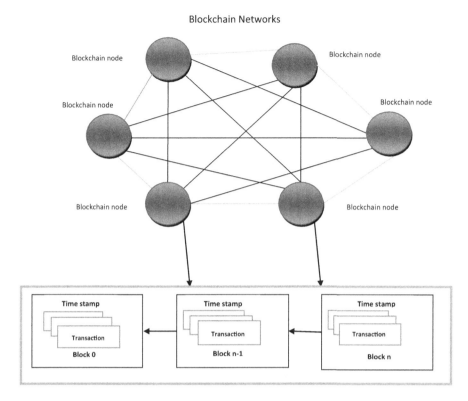

FIGURE 2.1   Blockchain architecture. Source: Authors.

criteria, peers must choose and approve these transactions. The nodes, for instance, attempt to confirm and authenticate by detecting if the initiator has enough money on hand to start a trade or by deceiving the system by making two purchases of these transactions. Spending twice is when money is used for two or more distinct transactions. The miners validate and confirm transaction, which is included in the block. A peer who mines for blocks using their computing power is referred to as a miner. To publish a block, miner nodes must use enough processing power to solve a mathematical challenge. It provides the option to make a whole new block. A bonus is given when a new block is successfully established. An agreement procedure is then used to have each peer in the network confirm the new block, a method used in decentralized networks to reach consensus. The latest block will then be produced locally, and each peer's immutable ledger will receive a copy that will be added to the current chain. The deal is finalized at this stage. The successive block adopts hash pointer for cryptography to link the newly generated block. The transaction is given a second confirmation, whereas the block is given a first confirmation. Additionally, the transaction will be reaffirmed whenever a new block extends the chain. Before a transaction is finished, six network confirmations are frequently needed (Murthy et al., 2020).

2.3 BLOCKCHAIN ARCHITECTURE

A small portion of work recorded in public documents might be called a blockchain transaction. Blocks are another name for these documents. All miners working on the blockchain network execute, put into place, and store these blocks for further confirmation. Evaluating past transactions is always possible, but updating them is not permitted. The blockchain, which is the underlying technology of Bitcoin, enables the decentralized global network of peer-to-peer exchanges. Bitcoin is, therefore, a censorship-resistant, universal form of electronic money. Trust may often be the primary problem in old centralized systems, such as banks, where users must rely entirely on the system. It is an area in which public blockchain technology excels since it allows ownership of digital products to be transferred from one peer to another without requiring trust. A trustless blockchain establishes trust by adhering to standards that disclose information about all network operations. Before beginning a transaction, security is another critical consideration. Consensus processes like blockchain mining, which are significantly dependent on cryptographic hash functions, can be utilized to solve security issues. As an illustration, Bitcoin uses the SHA-256 secure hashing algorithm. For inputs of any kind, the output of Bitcoin is 256 bits or 64 characters, known as a hash, which may be created from data of any kind, including text, numbers, strings, or even computer-created files of any size. The converted hash function output will always match the input. One-way functions are more names since it is impossible to compute the input from the output when the input is slightly changed while the output is altered. It is secure because one can only speculate about the input, and there are incredibly high probabilities that they are correct (Jabbar et al., 2022).

As the sender and not a third party requests the data exchange between sender and recipient, the first transaction stage procedure is to confirm sender's identity. Assuming that Bob and Alice each have a Bitcoin balance, Alice wishes to send Bob 10 Bitcoins, Alice will now publish transaction information on the network for sending the money. Private and public keys for digital signatures are employed by blockchain to do this. Bob's information includes transaction amount and public address which is accessible for the broadcast with Alice's public key and digital signature. Alice created that digital signature using her private key. All miners individually perform transaction validation based on several standards covered in more detail in this paragraph. In blockchain, the elliptic curve digital signature technique is employed. Thanks to this algorithm, the money can only be used by those who own it. Each transaction has a 256-bit signature, requiring 2,256 guesses to forge to generate a fraudulent transaction. An adversarial peer or attacker would be wasting their resources by trying to do this, which is computationally impossible. The validator must check both the sender's and the transaction's validity by assessing whether or not the sender has sufficient funds to deliver to the recipient. The ledger, which is filled with information about each preceding successful transaction, might be examined to carry it out.

2.3.1   BITCOIN TRANSACTIONS

The primary target of digital currency according to Bitcoin whitepaper is to eliminate central middlemen and establish a system of electronic money transfers between many parties that are decentralized. A Bitcoin transaction moves particular Bitcoins from one Bitcoin address to another. It usually starts in a client's Bitcoin wallet and is then broadcast to the network. Only if the transaction is genuine will the network nodes rebroadcast the block they are mining and incorporate it into it. Included in the transaction and other transactions in a block takes about 10 minutes. By this time, the receiver should be able to view the total transaction in their wallet. Unspent Transaction Output (UTXO) or the portion of a transaction's output that a user receives and has the option to spend in the future, is the primary component of a Bitcoin structure (Rajashekar et al., 2022). Assume that coins or cash in real wallet may become mixed up, although this differs from the received Bitcoin amount. The whole amount received and stored is stored in a Bitcoin wallet, for instance, if we maintain two sums ($2 and $3), they will total $5 in the same physical or digital wallet. As you are in the Bitcoin wallet, they will continue to be distinct entities and display the precise amounts. Consider Alice, who wants to give Bob 0.15 BTC but has only three distinct UTXOs in her wallet (0.01, 0.2, and 3). The wallet must choose one of these three output UTXO as a spend candidate to do that. If the wallet chooses to output 0.2, this amount will be unlocked, and the entire sum will be used in the UTXO entry field for the 0.15 BTC transactions. The next step is to transmit 0.15 BTC as an output UTXO to Bob's address wallet. The work required for miners to manage will be rewarded if they validate transaction by creating the fresh block and adding it to the existing chain.

$$\text{Transaction fee} = \text{Input} - \text{Output} \tag{2.1}$$

A successful miner is the one who collects the transaction fees and block creation incentives. Users frequently designate a transaction fee when sending transactions to the miners to reward them for successful block construction. The header will not contain information about transaction charge. Users may add transaction charge by delivering receivers a smaller sum than their total input UTXO. The value of this unassigned transaction can be viewed as a transaction charge. When mining a block, they also attempt to verify and examine the transaction data. Miners also include their coin-based transactions. A miner is the only person who may transact on coin-base, a particular form of Bitcoin transaction. This type of transaction generates a single output, which is all it has each time on the network; a fresh block is mined. The block reward, or compensation for a miner's work, is given during this transaction. Aside from that, the miner also sends any accumulated transaction fees. Before determining whether to add this record to the distributed ledger, the network peers decide on the leveling out of the transaction. The block reward and overall transaction fees will be delivered to the miner's provided address as part of the coin-base transaction. It illustrates the requirement that a miner distributes his reward when constructing a block. To ensure the block complies with the criterion, every node in the network will examine it, as described in Eq. (2.2). Transaction costs and the block reward are only usable by a miner once the block has been validated.

$$\text{sum(block\_output)} \leq \text{sum(block\_input)} + \text{reward} \qquad (2.2)$$

2.3.2 ETHEREUM TRANSACTIONS

The wallet program's reference implementation, UTXO, which maintained the account reference, is used to describe the current status of Bitcoin. The idea of an account was developed by Ethereum, a protocol component that acts as both the source and the destination of a transaction. Since the state is not preserved, as in Bitcoin UTXOs, the data, communications, and value transfers between the accounts via transactions permit state modification; accounts in Ethereum can be either externally owned accounts (EOA) or contract accounts (CA). Private keys belong to EOA, whereas codes govern CA, which can only be used with an EOA to activate it. While CA symbolizes an intelligent contract (SC), with the help of transactions, EOA communicates with the blockchain and is necessary to join the Ethereum network. On the blockchain's node, a bit of code called the SC extends the trust architecture by adding a layer of logic and processing. A message included in the transactions starts the execution of an SC.

The movable currency in Ethereum is called ether. Ether is denoted by the symbol. Fields for ether delivery and messages for launching intelligent contracts are both present in the Ethereum transaction. Transaction data, nonces, and previous block hashes are all properties that both Ethereum and Bitcoin share. It also uses other fields like the maximum charge and the state of South Carolina. The amount to transfer, the recipient address, and a simple ether transfer are made possible by providing the charges, gasoline rewards, and necessary accounts. Considering the nonce combination, the time stamp, and suitable execution fees, all produced transactions will be verified (Maalla et al., 2022). For the production of blocks, Ethereum employs an incentive-based methodology. For every operation, Ethereum needs cryptocurrency gas or fuel. It is used instead of ether as a charge for more straightforward computation. The fundamental justification for this is that gas is a cryptocurrency whose value is unrelated to transaction fees and processing costs. Unlike gas points, ether's value as a coin evolves in response to market shifts. Gas stations necessary for a transaction's completion are calculated throughout the mining process. Transactions using gas points are only allowed if the fee stipulated therein is sufficient. For the execution to occur, the necessary gas points must be present, both the intended transaction and the account balance. The originating account will receive any excess funds following the transaction's execution. The miners in Ethereum compete for block creation under its mining incentive mechanism. The miner who completes the riddle first is the winner; while the miner who completes later is an omer. Subsequent side blocks are included to the main chain as omer blocks after the winning block

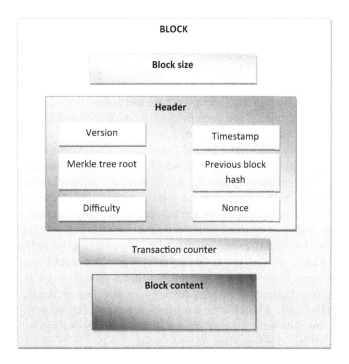

FIGURE 2.2 Block structure. Source: Authors.

is put into the main chain. The transaction fees are given to the winner block as gas points and the standard charge of three ethers. A little portion of the total gas points go to the ommer's block (Agrawal et al., 2022).

2.3.3 BLOCK STRUCTURE

A chain of blocks that make up the blockchain functions is comparable to a public ledger as a repository for every transaction. These blocks are connected by a reference hash that is either a parent block component or predecessor block. The phrase "first block" specifies the genesis block, unrelated to any other blocks (See Figure 2.2). A block header and a block of content make up a block. Block version, Merkle tree root hash, parent block hash, nBits, time stamp, and nonce are a few examples of hash values that may be found in the block header and other metadata.

2.4 BLOCKCHAIN CHARACTERISTICS

Decentralization: In conventional centralized transaction system the central trusted agency (central bank) must first authorize every transaction. Blockchain technology with a decentralized peer-to-peer structure is a favorable option since decentralization necessitates trust, which is the major problem, as well as boosting robustness, availability, and failover. In contrast to centralized systems, any two peers (P2P) can conduct a transaction on the blockchain network without the need for central-ized authentication. By employing several consensus techniques, blockchain might lessen the problem of trust. Additionally, it can lower server expenses (such as setup and maintenance costs) and loosen performance constraints at the primary server. Blockchain, in contrast, typically has certain disadvantages. For instance, in PoW environments like Bitcoin and Ethereum, the cost of servers and electricity has increased by several orders, and performance has declined significantly. The block-chain characteristics are shown in Figure 2.3.

Persistency: Both data producers and consumers can prove the quality and integ-rity of their data thanks to blockchain, which provides the foundation for assess-ing the authenticity of information. As an illustration, if a Blockchain includes ten blocks, the block's hash following block ten is contained in block number 10, and new block is made using data from the current block. The outcome is that every block is connected to every other. Even the transactions themselves have connections to earlier ones. A straightforward modification to any transaction will dramatically alter the block hash. All the hash information from previous blocks must be changed if somebody wants to modify any information that is thought to be astronomically impossible, given the quantity of work required, in an endeavor. Additionally, other network users validate a block once a miner generates it. The network will be able to recognize any data fabrication or change. As a result, blockchain is viewed as an immutable distributed ledger and is tamper-proof.

Anonymity: It may connect to blockchain network using randomly produced address. In a blockchain network, an individual may employ many addresses to con-ceal his identity. Owing to the decentralized system nature, no one in control is monitoring or gathering the personal information of its users. Because of its distrust-less environment, blockchain offers a certain amount of anonymity.

Auditability: A digital time stamp and a distributed ledger record and verify every transaction inside a blockchain network. Each network node may be accessed. As a result, historical data may be audited and tracked. For instance, Bitcoin allows for the iterative tracing of all transactions, helping to make the data state on the

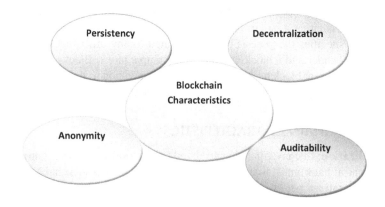

FIGURE 2.3 Blockchain characteristics. Source: Authors.

blockchain more transparent and auditable. However, it becomes exceedingly challenging to track down the source of the money when it is spread over several accounts.

2.5 BLOCKCHAIN TAXONOMY

Consortia, public, and private blockchains are the three different kinds as in Figure 2.4. These systems may be contrasted from several angles as discussed below.

Determining Consensus: In a public blockchain like Bitcoin, while every node can participate in the consensus procedure, only a select few carefully picked nodes are accountable for block validation in a consortium blockchain. A central decision-maker will choose the delegates in a private blockchain who will select the block that has been validated.

Read Permission: Users on public blockchains can read data. However, those on private and consortium-distributed ledgers can only have limited access. Thus, the consortium/organization can choose whether the stored information must remain accessible to the general public.

Immutability: Changing the public blockchain is challenging since every transaction is validated by every peer and recorded in the distributed ledger on decentralized blockchain network. Contrarily, private blockchain consortium and ledger are subject to alteration at the whim of the in-charge.

Effectiveness: Any node may join or exit the network at any time, thus enhancing the scalability of the public blockchain. However, the network's adaptability to additional nodes and the growing complexity of the mining process have a restricted throughput and rising latency. However, private and consortium blockchains can use fewer validators and optional consensus techniques, increasing performance and energy efficiency.

Centralized: One entity has power over the private blockchain, the consortium is relatively concentrated, and the public blockchain is decentralized, distinguishing these three types of blockchain considerably from one another. Many consumers may be drawn to public blockchain since it is accessible to

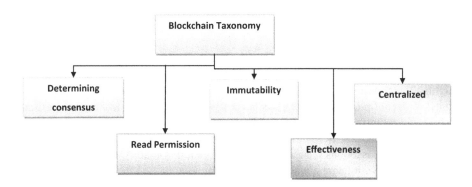

FIGURE 2.4   Blockchain taxonomy. Source: Authors.

everyone. Communities are highly active as well. New public blockchains start to appear on a daily basis. There are several commercial uses for consortium blockchain. Hyperledger is presently designing blockchain frameworks for corporate consortiums. Instruments for creating consortium blockchains have also been made available by Ethereum. For efficiency and auditability, many businesses continue to use the private blockchain.

2.6 DIGITAL CURRENCIES

A kind of currency known as a "digital currency" has no physical counterparts, such as coins or paper money, and only exists digitally. They may be electronically transmitted between parties and are often used for internet transactions (Xie et al., 2020). Digital currencies are becoming more and more widely used by individuals and companies throughout the world. There are two types of digital currencies: Cryptocurrency and Central Bank Digital Currency (CBDC).

2.6.1 CENTRALIZED BANK DIGITAL CURRENCIES

CBDCs are electronic analogues of traditional fiat currencies that central banks issue and support. CBDCs may be used for several purposes, including minimizing cash management costs, increasing financial inclusion, and simplifying cross-border payments. For CBDC, using blockchain technology is not required; it is an alternative. Several technology approaches may be used to implement CBDCs. With blockchain technologies distributed and transparent ledger, one way to create CBDCs is to trace transactions to ensure the currency's integrity. The Central Bahamas Bank is made of Sand Money which is a blockchain-based digital alternative to the Bahamian currency (Wamba et al., 2020), and other banks are now assessing the adopting blockchain technological potential for CBDCs. Other central banks, however, are looking into different strategies for putting CBDCs into place, using centralized databases or a system that blends components of both centralized and decentralized systems. The CBDC's specific goals, the financial system in place at the time, and several criteria will determine the technology method employed, including the legal environment in which it perform business.

2.6.2 CRYPTOCURRENCY

Virtual or digital currencies are often known as cryptocurrencies, which use cryptographic techniques for transaction security and to manage the generation of new units. A decentralized distributed ledger known as blockchain technology permanently and independently verifies transactions and is the foundation of most cryptocurrencies. Each of the interlinked blocks that make up a blockchain contains a collection of transactions related to the one before it in a chain-like structure (Atlam et al., 2018). A significant advancement in technology and finance is the emergence of currencies based on blockchains. A network of users validates transactions, and in exchange for their work, they are given fresh Bitcoin units. Because they are

TABLE 2.1
Various Currencies Comparision

Electronic Currencies	CBDCs	Cryptocurrencies
Issuance	Issued and approved by the monetary authority.	Generated using a method known as "mining."
Regulation	Must be managed at the same level as conventional fiat currencies.	Typically less controlled.
Decentralization	May employ a personal blockchain.	Operate in a decentralized network on a public blockchain.
Use examples	Value exchange and storage.	A platform for decentralized apps, a business, a store of money, etc.

decentralized and not under the jurisdiction of a single government or financial institution, comparable amounts of regulation do not apply to cryptocurrencies as they do to CBDCs or traditional fiat money. Transactions are secure using public and private key cryptography. Private keys are utilized to access and control a user's Bitcoin assets, while public keys are required to receive payments. In "mining," which involves users solving challenging mathematical puzzles to authenticate transactions, new Bitcoin units are produced and uploaded to the blockchain. Some cryptocurrencies need specialist gear (such as a powerful computer with a high-end GPU) to mine them because the procedure needs huge computing power and is computationally complicated (Zheng et al., 2018).

Cryptocurrencies employs consensus methods to verify transactions and keep the blockchain safe from hacking. Consensus processes include proof of stake (PoS), proof of work (PoW), and delegated proof of stake (DPoS). Every method has its own benefits and drawbacks and the selected mode significantly influences network's efficiency, expandability, and security. With varying degrees of convenience and security, users may manage, store, and transfer their Bitcoin holdings using cryptocurrency wallets (Joshi et al., 2018). Frequently used as a medium of exchange, cryptocurrencies may be purchased, sold, and exchanged peer to peer or on online exchanges. There are several distinct cryptocurrencies as shown in Table 2.1; each has its own special technological requirements and capabilities.

2.7 CRYPTOCURRENCY FUNCTIONALITY

The whole population, including those who lack access to traditional financial services, can utilize the e-rupee (e₹) (also known as the digital rupee); it aspires to deliver an easy, secure, and valuable payment system. It is anticipated to decrease the economy's reliance on cash by boosting digital payments. Many individuals think that e₹ is the currency of India's cryptocurrency. However, there are some

TABLE 2.2
Types of Rupee

Token-Based	Account-Based
It is shown on a blockchain as a digital token.	It does not appear on a blockchain as a digital token.
A distinct digital asset may be traded between individuals through transactions.	It is a balance linked to a user's account rather than a single digital asset.
It may be traded for fiat money or other cryptocurrencies.	It is solely usable inside the system in which it is issued and cannot be traded for other money.

differences between the e₹ and cryptocurrencies. The digital rupee would function similar to CBDCs because it digitally represents real money (Dave et al., 2019). It would be designed as a wealth store and commerce medium like conventional fiat currencies. The same coin and paper currency values will be used to distribute the e₹ via the intermediary banks. The official website, the mobile app, and reputable banks allow users to buy digital rupees. It's crucial to remember that the digital rupee is still a young currency and very little is understood about how it works (See Table 2.2).

(i) **Direct Model**: The RBI is the central processing entity for all transactions in this approach. The central authorities supervise e₹ issuance, transaction monitoring, and supply management. The single-layer model is another name for it.

(ii) **Two-Tier Model**: The indirect model is another name for it. In this concept, a group of scattered nodes are controlled by a single entity (RBI) working together to process every transaction. The decentralized nodes are accountable for validating and logging transactions on a ledger, while the central authority produces the e₹ and keeps track of all transactions (Monrat et al., 2019).

(iii) **Hybrid Model**: A variant of the e₹ design known as the hybrid model contains components from models with one tier and two tiers. A system of nodes with no central authority is in charge of validating and logging transactions on the ledger. In contrast, the RBI, a main body, is responsible for issuing electronic rupees and keeping track of all transactions. The single-and two-tier models' benefits are both still there in the hybrid model. However, they are each meant to be lessened while still being beneficial. The flow shows the main steps for creating and dispersing a digital currency provided by a central bank that supports the Reserve Bank of India (RBI). Many technologies could be utilized, depending on how the specific CBDC is designed and implemented; nevertheless, some often-used technologies might include:
Distributed Ledger Technology (DLT): In DLT systems, CBDCs may be created to provide the safe and decentralized recording of transactions.

Smart Contracts: Considering that they are self-executing contracts whose terms are expressly set down in the code, they may be used as legal instruments by CBDCs. Process automation and transaction costs can both be increased as a result.

Cryptography: Cryptography is a tool CBDCs may use to safeguard transactions and stop fraud. Digital signatures, hashing, and encryption may guarantee transaction secrecy and authenticity.

Application Programming Interfaces (APIs): APIs may be used by CBDCs to connect to the infrastructure and payment systems already in place. Interoperability may be improved, and smooth transactions between various payment systems may be possible.

Mobile Wallets: Downloadable mobile applications that function as mobile wallets can be used to store and deal with CBDCs. An accessible and user-friendly interface for CBDC transactions may be offered via mobile wallets.

Digital Identity: As digital identity providers, CBDCs may be connected to biometric or national identification systems. It may be beneficial to guarantee the authenticity and security of CBDC transactions. The technologies employed by CBDCs are created to offer electronic money that can be used for routine transactions that are safe, dependable, and readily available. CBDCs can provide various benefits over traditional fiat currencies by utilizing cutting-edge technologies; among them are increased security, cheaper transaction costs, faster transaction speeds, and a more comprehensive range of financial inclusion (Shobanadevi et al., 2022).

2.8 INDIAN RUPEE SYSTEM

An e-rupee system's specific design will be determined by several elements, including its objectives and specifications, technological prowess, and the societal, political, and legal framework in which it will operate. A Private Blockchain network will power all regulated banks, Reserve Bank and transactions and other licensed banks' transactions with account holders, customers, companies, and other banks with licenses will be powered by a Consortium Blockchain, with or without the assistance of third-party apps. The two blockchain network type designed for exclusive, restricted access are private and consortium.

2.8.1 PRIVATE BLOCKCHAIN

Even though it is anticipated that regulated banks and reserve banks would still require a more efficient private blockchain network that can process many transactions, a private blockchain allows transactions to be handled quickly and efficiently,. Since its nodes are controlled by a single business, using a private blockchain reduces the number of nodes required to attain consensus; the consensus process may be cranked up for speed. The recommended method calls for incorporating Income Tax

Authority, SEBI, CAB, and Enforcement Directorate regulatory compliance organizations into a personal blockchain for monitoring and verifying transactions. Using CBDC as a private blockchain is possible (Hazra et al., 2022).

2.8.2 CONSORTIUM BLOCKCHAIN

A decentralized network is created when many companies or entities join forces to create a consortium blockchain. This kind of network is appropriate for digital money; referred to as Central Bank. Central banks developed digital currency (CBDC) since it successfully balances scalability, security, and anonymity. In addition, in the context of the Indian economy, constructing a consortium blockchain network would offer a secure, private network with many regulated organizations and third-party apps. All digital apps must have built-in operating compatibility for transactions involving the CBDC ande carried out by NEFT, RTGS, UPI, or quick money transfer. It is displayed for both the consortium blockchain and the private blockchain. In contrast to open blockchain technology, private and consortium blockchains provide several benefits, such as:

Privacy: Due to network access being limited to approved users, compared to public blockchains, private and consortium blockchains provide more privacy. It may be particularly relevant to organizations that handle confidential or sensitive data.

Scalability: Because they don't need as much computing power to operate the network, public blockchains may be more challenging to grow than private and consortium blockchains. They may be more economical and practical for particular kinds of transactions (Yu et al., 2020).

Governance: Given that they are controlled by a single business or a collaboration of several companies, private and consortium blockchains provide more network control. Increased accountability may result from improved ease of implementing network updates and modifications.

Flexibility: Applications for private or consortium blockchains is customized to fulfill the working environment or consortiums' unique needs. In contrast to public blockchains, it may not be capable to assist all sorts of use cases and transactions, and private blockchains are more flexible.

Compliance: Consortium and private blockchains may be helpful for companies working in highly regulated sectors like banking or healthcare since they may be built to meet specific regulatory standards (Rathod et al., 2022).

2.9 ANALYSIS

The Indian rupee may be shown in a digital format, according to RBI. Based on the facts now available and the broad trends, theoretical and practical feasibility of digital currencies may yet be discussed.

2.9.1 THEORETICAL ANALYSIS

Theoretically, an Indian digital currency might offer several advantages, such as:

Increased Financial Inclusion: In India, a sizable portion of the population lacks access to standard financial services and is unbanked. More widespread financial inclusion may be made possible by a digital currency that offers these people a cheap and convenient way to exchange value.

Reduced Transaction Costs: By doing away with intermediaries like banks and payment processors, digital currencies can minimize transaction costs, increasing efficiency and lowering prices for consumers and companies.

Improved Transparency: Financial transactions might become more transparent thanks to digital currencies, making monitoring cash movements more straightforward and stopping illegal practices like money laundering.

2.9.2 IMPLEMENTATION

An extensive and dependable technological infrastructure and regulatory backing that can manage high transaction volumes and guarantee the implementation of digital currency must be reliable in an Indian digital rupee system. The following are some of the significant variables that might determine whether or not a performance is successful:

Technical Infrastructure: For a significant digital money system to work, a strong technological foundation is needed to handle the volume and speed of transactions. It would necessitate a substantial investment in hardware, software, and cybersecurity measures to guarantee the system's safety and integrity.

Regulatory Framework: A favorable legal and regulatory environment is necessary and established for digital currencies to function. Any initiative using digital coins would have to work inside the intricate legal system that governs financial services in India while also considering problems like secure and private data.

Public Acceptance: Any initiative involving digital money must be adopted and accepted by the general public to be successful. Any industry using digital money would need to consider India's varied population and uneven levels of technical knowledge to secure universal adoption.

Finally, the practicality of its implementation would depend on several variables despite the obvious theoretical advantages of digital money in India, including its technological infrastructure, legal system, and level of public acceptability. The RBI may or may not move through with a digital currency project. However, any business of this nature would need a substantial investment, careful planning, and governmental permission.

2.10   EVALUATION WITH ANOTHER CENTRALIZED BANK SYSTEM

CBDCs are intended to supplement or replace real cash in the economy by offering a safe, effective, and dependable way to exchange digital currency. Several nations were considering introducing central bank digital currencies (CBDCs). Digital currency electronic payment (DCEP), one of the CBDCs, is a service provided by People's Bank of China that is currently being developed the most. The Sand Dollar, Central Bahamas' Bank first CBDC, was released in 2020. The establishment of an electronic krona is being considered by Riksbank, Sweden's central bank, with consumers anticipating saving in a digital wallet, which would function as an account-based system (Jyoti et al., 2022). The future development and separation of these systems from one another will be fascinating. However, comparing the characteristics of different CBDCs can be difficult because every nation would have a separate set of particular conditions, requirements, and goals. Some of the critical distinctions between the e-rupee and the CBDCs are discussed in Table 2.3.

2.11   PROSPECTS WITH THE INDIAN ECONOMY

2.11.1   CENTRALIZED

The Indian government has made efforts to foster the growth and usage of the e-rupee and has shown interest in encouraging the use of virtual money. It may promote the adoption of e-rupee and boost consumer confidence in it. The government will accept digital rupees as a legitimate currency. Unlike other cryptocurrencies, RBI rather than being in charge of digital rupee will have some control over it. Any transaction on approved networks is accessible to the Reserve Bank of India and the executive branch.

2.11.2   SECURE

By utilizing cutting-edge cryptographic algorithms, a digital rupee could offer higher security than traditional physical money thanks to decreased vulnerability to physical theft, decentralized ledger technology, and multifactor authentication. Aadhaar, an Indian biometric identity system, may also be connected to it, allowing consumers to accept payments straight, without needing physical signatures or written verification, into their bank accounts. A digital currency will have an indefinite lifespan compared to real money because it cannot be physically harmed or lost. In contrast to actual cash, digital money leaves a trail of data that is simpler to follow and validate. It can aid the discovery and avoidance of fraud, money laundering, and other unlawful activities. It is thus predicted that the digital rupee would have a robust security mechanism that uses encryption and a consensus procedure to prevent duplicate expenditures and other nefarious actions.

TABLE 2.3
Comparison of Various Rupee Systems

CBDC	e-Rupee	Sand Dollar	E-Krona	DCEP
Release Date	On December, 2022, a pilot experiment was launched.	October, 2020	The project is being developed and tested; no specific launch date exists.	Trial proceedings began in April 2020.
Technology	Blockchain technology with a centralized hybrid approach.	A system with two tiers that is based on blockchain and is centrally managed.	A system with a two-tier architecture that is built on a centralized blockchain.	A system with a centralized two-tier architecture and permission on a blockchain.
Payment Mechanism	Pay-as-you-go digital money.	Fiat money in digital form.	Fiat money in digital form.	Pay-as-you-go digital money.
Accessibility	Everyone with a smartphone can access.	Only accessible to Bahamas citizens.	expected to be solely available to Swedish citizens.	Accessible to everyone who owns a smartphone.
Interoperability	Incompatible with other digital currencies.	Interoperability with various digital currencies is a design goal.	We anticipate working with more blockchain-based currencies.	Unable to work with other coins.
Offline Capabilities	There has to be an offline function.	Connectivity to the internet is required.	Connectivity to the internet is required.	Has the ability for offline payments.
Privacy	Continuing to assess privacy concerns.	Use a method called zero-knowledge proofs to increase privacy.	Continuing to evaluate privacy concerns.	Since the government has states that DCEP transactions can be traced, there are significant privacy concerns.

(Continued)

TABLE 2.3 (CONTINUED)
Comparison of Various Rupee Systems

CBDC	e-Rupee	Sand Dollar	E-Krona	DCEP
Regional Coverage	Not certain.	Interior of the Bahamas.	Unique to Sweden.	Designed for use both domestically and internationally.
Status	Testing is still ongoing.	Completely functional.	Testing is still continuous.	Implemented in several major cities, including Beijing, Chengdu, and Shanghai.

2.11.3 EASE OF USE

Without a real bank or requirement for actual currency, digital currency may be utilized whenever and anywhere. Particularly for individuals who live in isolated or rural areas, this can increase the convenience and effectiveness of transactions. To use e-rupee, bank account isn't necessary. The bank may yet let us purchase digital rupees as tokens. Generally, it is similar to withdrawing money from banks, which would credit our electronic wallets with funds from our bank accounts rather than delivering us cash, letting us use regular currency. Account payouts will be real-time with the transactions made online.

2.11.4 GLOBAL ACCEPTANCE

Cross-border currency conversions and money transfers take time and money. Digital rupee will enhance bank cash management and operations through quick cross-border transfers. No matter where they are, the digital rupee is used for international financial transactions by NRIs who own it. The expansion of Indian economic initiatives will be aided by it (Li et al., 2022).

2.11.5 FAVORABLE EFFECT ON THE ECONOMY

With a ratio of 17% to GDP, India has a more significant propensity to withdraw cash than Nordic countries like Australia and the UK. If people utilize e-rupee, they could become less reliant on money. Naturally, it will lower the costs of administering, printing, and dispersing actual money. Additionally, using a digital rupee can aid in reducing the volume of illegal financial exchanges and uses, which can boost tax receipts and lessen corruption. In addition to giving the government access to additional data and information, the development of digital money brings us to our final point about consumer purchasing trends, which they can use to enhance the efficiency of economic programs and increase income. The introduction of a digital rupee might, all things considered, have a perfect effect on the Indian economy over time and provide tax money. The primary trend in the current era of electronic payment is the expansion of digital money. Due to the numerous advantages of the CBDC, more and more countries are researching it. Digital payment methods have become much more prevalent in India and this tendency is projected to endure. It may result in a desire for the e-rupee as a substitute for other digital payment methods (Xiong et al., 2022).

2.11.6 INSURANCE

Transactions between multiple customers, policyholders, and insurance firms can be supported by blockchain in the insurance sector. As well as supporting reinsurance transactions between insurance firms, blockchain may be used to bargain, purchase, and register insurance policies. Smart contracts may be used to automate different insurance plans, which can drastically save administrative expenses. Processing insurance claims, for instance, entails a significant administrative expense. Because of misunderstandings and varying interpretations of the words, the administration of claims can sometimes be a highly complicated procedure. Smart contracts can prevent these issues by adequately describing the if–then interactions between insurance policies. Automating the term execution through digital protocols that correctly perform agreed-upon insurance policies used can reduce insurance cost and work associated with implementation. With this decrease, insurance firms can diminish the insurance policies' cost and increase their competitiveness to attract more clients. At the same time, it makes it possible for insurance companies to provide their clients with brand-new automated insurance options without having any issues about administrative costs. Blockchain also makes it possible for insurance businesses to grow internationally (Zhang et al., 2022).

2.11.7 FINANCE

A letter of credit (LC), which is helpful for risk mitigation, is used by banks to expedite the trade financing process. Although it still makes up more than one-fifth of global commerce, the procedure is complicated, expensive, and subject to contractual delays. The trading parties find it less appealing to conduct low-value transactions when the time and expense associated with issuing LC grow. In addition to eliminating banks, this trend promotes free trade. Blockchain may address these issues by streamlining processes and reducing transaction costs by automating LC. As stated in the LC, the conditions required by the supplier and client may be included in the blockchain's smart contract architecture to assure payment when the traded products are delivered to the customer. This strategy may decrease the time and expense associated with LC changes by reducing contractual ambiguities and informational disparities. Blockchain has the potential to be the answer to accelerating the documentation process while maintaining security; even though the ICC poll revealed that over 80% of respondents were concerned about conventional trade finance, they predicted that it would not change or grow shortly (Chung et al., 2016).

2.12 CHALLENGES

Blockchain is currently one of the terms used most frequently in business and technology. The financial sector is predicted to undergo a significant transformation because of this technology since it can function without the aid of a centralized authority or intermediary. Due to its ability to keep a vast trail of records and store tamper-proof data, blockchain is also anticipated to be advantageous for other industries. Blockchain has limitations (See Figure 2.5), though, and many business models are separate from it, much like other new technology.

2.12.1 DIGITAL ILLITERACY

In 2021, India was rated 73rd out of 120 nations regarding digital literacy. The primary cause is that high-speed internet access still needs to be available in many rural

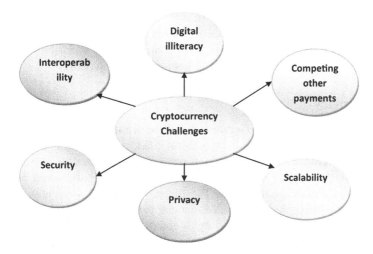

FIGURE 2.5 Cryptocurrency challenges. Source: Authors.

parts of India. Therefore, people in those places encounter difficulties in utilizing the benefits of the digital revolution. India must find a solution to this problem to achieve its goal of promoting digital currency.

2.12.2 COMPETING WITH OTHER PAYMENT METHODS

e-Rupee can compete with other digital payment choices, including existing cryptocurrencies and bank-based digital payment systems, regarding usability, supporting infrastructure, creative mechanisms, and reasonable transaction prices. The people of India are multilingual and speak several different languages. The architecture should support many languages and offer a simple user interface for users who must be fluent in English. Due to the price-conscious nature of the Indian market, excessive transaction fees are likely to repel clients. The design should provide affordable transaction rates to promote adoption and usage. Incentives might be used to encourage the usage of digital currencies because there are many unbanked and under-banked people in India. Strategies like referral or transaction rewards should be part of the design.

2.12.3 SCALABILITY

Blockchain and cryptocurrency-based business model solutions are increasing across all industries. Its scalability and performance, in particular, make one wonder if it meet the expanding demand from diverse governmental and business sectors. Some researchers concentrate on performance issues like throughput (number of transactions/second), latency (total time needed to add transaction block to the blockchain), and scalability concerns on number of network replicas. Performance and latency may suffer if more copies are added because the network must manage more messages being exchanged and processed. PoW is one of the protocols that can guarantee scalability, even if it has a high latency and limited throughput. This bottleneck is caused by the resources needed to post a block for inclusion in the chain, which requires solving a cryptocurrency problem (Hoy et al., 2017).

2.12.4 PRIVACY

Blockchain is designed to give protection and anonymity to delicate personal data, since transactions may be done using generated addresses rather than real names. Even if other network users may, several researchers theorized that as others verify the public key employed to initiate transaction, blockchain would be vulnerable regarding transactional privacy. The Bitcoin platform has recently undergone research, and it has been demonstrated that comparing a member's transaction history might reveal their genuine identity. It contrasts the claim that peers can preserve anonymity within the lockchain network. Peck et al. (2016) proposed that a technique for connecting IP addresses can be converted from peers' aliases even if NAT or firewalls protect them. He claims that a peer's linked group of nodes, which make up that peer, might be applied to identify that peer specifically. The equality of all balances is the main factor contributing to block chain's sensitivity to information leaking, and data associated with public keys are available to all users on the network. In light of this, the privacy and security criteria for blockchain applications should be established at the outset.

2.12.5 INTEROPERABILITY

Numerous sectors may be interested in using blockchain technology at the moment. But they can't cooperate and integrate since there isn't a set mechanism. This problem is referred to as a lack of interoperability and is harmful to the development of the blockchain sector. As a result, rather than offering diverse real-world solutions to various business models, Bitcoin still serves as the primary platform for blockchain technology. Blockchain programmers may build code for multiple platforms because they have no interoperability, but each network is locked off from the others and cannot connect. For instance, on GitHub, over 6,500 blockchain projects are active using various platforms, protocols, consensus procedures, computer languages, and security measures. Because of this, standardization is needed for corporate cooperation in application development to distribute blockchain-based products and incorporate them into current platforms.

2.12.6 ENERGY

The proof-of-work (PoW) algorithm has enabled Bitcoin to facilitate P2P transactions in potentially dispersed, decentralized environment. Moreover, miner computers require huge energy to perform the activity. Bitcoin's force usage indicator demonstrates how seriously unsustainable the PoW algorithm is. The system of incentives encourages individuals to mine Bitcoin all around the world. Since mining produces a consistent flow of revenue, people are driven to run power-hungry equipment to participate in the industry. The cost of Bitcoin increased, as did the network's average energy consumption rate.

2.12.7 SECURITY

The most widely acknowledged issue with blockchain security is 51% of assaults. One or more malevolent actors take over most of a blockchain's hash rate during a 51% assault. With the hash rate, it may commit double-spending by reversing transactions and preventing blocks from being validated by other miners. The unethical practice of "selfish mining", utilized by mining pools to increase block rewards, affects a blockchain network's reliability. If anyone wishes to use hashing power, the network is still susceptible to fraud even though it is typically considered that rogue nodes with more than 51% of the computational blockchain network may come under the grip of power. A miner or groups of miners can inductee the selfish mining process by electing not to broadcast the validated blocks to the available network. They continued mining for the following block to maintain their edge. Only after specific requirements have been satisfied may the general public view the solved blocks. Because of this, the self-centered network is compelled to accept a miner's solutions as their chain grows more prolonged and complex, while other miners expend energy on a useless branch. At last, rapacious miners want more cash that makes it more likely to selfish pool with more than 51% of the power since it is tempted to join the longer chain, the logical miners (Tschorsch et al., 2016).

2.13 FUTURE RESEARCH WITH BITCOIN

Researchers assert that blockchain technology has great promise for academic and commercial applications. Standardization, asset protection, big data, and smart contracts are only a few of the many potential applications for blockchain technology that have been briefly covered in this section. Investors can be seduced by the prospect of enormous profits when using blockchain technology. To properly implement this technology in a commercial solution, it is essential to ascertain whether the criteria are met. It is crucial to examine the advantages and disadvantages of blockchain-based solutions using a systematic testing process. The standardization and testing phases are the two different stages of this procedure. The first step will evaluate the developers' claims on their blockchain solutions based on a specified set of criteria. The blockchain-based system's functioning is assessed during the testing phase. For instance, the owner of an online retailer is interested in how the blockchain-based system works. The platform used for the purchased solution should have certain testing and standardization approaches to assess its capacity, throughput and latency (Yli-Huumo et al., 2016).

By registering new ideas, proofs of concept, and drawings using blockchain technology, businesses may build a digital trail of their creations and produce certificates that can be used to demonstrate the legitimacy, existence, and ownership of any intellectual property asset. All notarized information, including trade secrets and copyright claims, might be kept private and safe using the special cryptography layer. In particular, big data analytics and blockchain might work effectively together for data management and analytics. Blockchain technology may be used to manage data by storing it securely and decentralized (Zheng et al., 2018). Additionally, the immutability aspect of blockchain could guarantee the data's validity. As an illustration, consider how difficult it would be to change or steal patient health information stored in a distributed ledger without the owner's consent. Data analytics may be applied to blockchain transactions. This approach makes learning more about potential business partners' trade habits and online conduct feasible.

2.14 CONCLUSION

The potential uses, advantages, and restrictions of the blockchain are evaluated in this research via a wider analysis. The consensus mechanisms, application areas, system architecture, and blockchain transaction process are also fully described. Several challenges still require additional research and analysis to develop more successful industrial applications that fully utilize blockchain technology and achieve the intended goals in Indian economy. India is a country where a considerable portion of transactions are made using actual money, making it a civilization that relies heavily on cash. It may be an issue since it would be expensive and time-consuming to print and distribute genuine money. There are chances that it would be counterfeited too, and it would be challenging to keep track of tax transactions. RBI's e-initiative targets to replace physical cash in wallets as its primary objective. Use QR codes or parties' digital rupee wallets to send and receive payments. Because of this, it could be simpler to perform electronic transactions and payments might trigger financial inclusion and promote economic growth. Payments are accepted in e-rupee, which may be exchanged for current currencies and kept securely. The new payment method's slow acceptance by the public and conventional financial institutions makes adoption difficult. Accepting the e-rupee as a mode of payment by businesses is a requirement for the currency's success. It will be challenging for the e-Rupee to achieve widespread adoption if retailers are unwilling to accept it. Awareness-raising initiatives and education are required to overcome these obstacles and encourage adoption.

The design can communicate with widely used payment methods in India including UPI and digital wallets, to convert digital currency for fiat money as simply as possible. To increase public confidence, the government should establish clear regulations. In addition to ensuring that the required infrastructure is in place to facilitate the usage of a digital currency, it is crucial to ensure that public is informed regarding its uses and potential threats. Money security will always be a significant challenge via cryptographic protocols and other measures. Also, efforts must be taken to increase the network's security. Using digital currencies in India presents particular challenges. Some problems need a long-term plan, even if many may be resolved quickly. India has a fantastic opportunity to lead the globe because of the extensive adoption of digital currency. The e-rupee may be an innovation accelerator by encouraging competition and efficient payments. Hopefully, this endeavor will spark further conversation on the appropriate action.

REFERENCES

Atlam, H.F.; Alenezi, A.; Alassafi, M.O.; Wills, G. Blockchain with the Internet of things: Benefits, challenges, and future directions. *Int. J. Intell. Syst. Appl.* 2018, 10, 40–48.

Agrawal, D.; Minocha, S.; Namasudra, S.; Gandomi, A.H. A robust drug recall supply chain management system using hyper ledger blockchain ecosystem. *Comput. Biol. Med.* 2022, 140, 105100.

Chung, M.; Kim, J. The internet information and technology research directions based on the fourth industrial revolution. *KSII Trans. Internet Inf. Syst.* 2016, 10(3), 1311–1320.

Dave, D.; Parikh, S.; Patel, R.; Doshi, N. A survey on blockchain technology and its proposed solutions. *Procedia Comput. Sci.* 2019, 160, 740–745.

Hazra, A.; Alkhayyat, A.; Adhikari, M. Blockchain-aided integrated edge framework of cybersecurity for internet of things. *IEEE Consum. Electron. Mag.* 2022. https://doi .org/10.1109/mce.2022.3141068

Hoy, M.B., An introduction to the blockchain and its implications for libraries and medicine. *Med. Ref. Serv. Q.* 2017, 36(3), 273–279.

Jyoti, A.; Chauhan, R. A blockchain and intelligent contract-based data provenance collection and storing in the cloud environment. *Wirel. Netw.* 2022, 28, 1541–1562.

Joshi, A.P.; Han, M.; Wang, Y. A survey on security and privacy issues of blockchain technology. *Math. Found. Comput.* 2018, 1, 121.

Jabbar, R.; Dhib, E.; ben Said, A.; Krichen, M.; Fetais, N.; Zaidan, E.; Barkaoui, K. Blockchain technology for intelligent transportation systems: A systematic literature review. *IEEE Access.* 2022, 10, 20995–21031.

Kaur, H.; Alam, M.A.; Jameel, R.; Mourya, A.K.; Chang, V. A proposed solution and future direction for blockchain-based heterogeneous medicare data in a cloud environment. *J. Med. Syst.* 2018, 42, 156.

Kili Çarslan, S.K. Bitcoin Özelİnde Kripto paralarin Edinilmiş mallara katilma rejiminde tasfiyesi sorunu. *Kırıkkale Hukuk Mecmuası.* 2023, 3, 1–27.

Li, S.; Zhang, Y.; Xu, C.; Cheng, N.; Liu, Z.; Du, Y.; Shen, X. HealthFort: A cloud-based ehealth system with conditional forward transparency and secure provenance via blockchain. *IEEE Trans. Mob. Comput.* 2022, 1–18. https://doi.org/10.1109/tmc.2022 .3199048

Maalla, M.A.; Bezzateev, S.V. Efficient incremental hash chain with the probabilistic filter-based method to update blockchain light nodes. *Sci. Tech. J. Inf. Technol. Mech. Opt.* 2022, 22, 538–546.

Mougayar, W. *The Business Blockchain: Promise, Practice, and Application of the Next Internet Technology.* John Wiley & Sons: Hoboken, NJ, USA, 2016.

Murthy, C.V.B.; Shri, M.L. A survey on integrating cloud computing with blockchain. In *Proceedings of the 2020 International Conference on Emerging Trends in Information Technology and Engineering (ic-ETITE)*, Vellore, India, 24–25 February 2020, pp. 1–6.

Monrat, A.A.; Schelén, O.; Andersson, K. A blockchain survey from the perspectives of applications, challenges, and opportunities. *IEEE Access.* 2019, 7, 117134–117151.

Peck, M. A blockchain currency that beat Bitcoin on privacy. *IEEE Spectrum.* 2016, 53(12), 11–13.

Rajashekar, M.; Sundaram, S. Dynamic attribute tree for the data encryption and third party auditing for cloud storage. *Indian J. Sci. Technol.* 2022, 15, 798–805.

Rathod, T.; Jadav, N.K.; Alshehri, M.D.; Tanwar, S.; Sharma, R.; Felseghi, R.A.; Raboaca, M.S. Blockchain for future wireless networks: A decade survey. *Sensors.* 2022, 22, 418.

Strilets, B. Current state and prospects for the legal regulation of cryptocurrencies in the European Union. *Actual Probl. Law.* 2022, 2, 70–76.

Tschorsch, F.; Scheuermann, B. Bitcoin and beyond A technical survey on decentralized digital currencies. *IEEE Commun. Surv. Tutorials.* 2016 18(3), 2084–2123.

Wamba, S.F; Queiroz, M.M. Blockchain in the operations and supply chain management: Benefits, challenges and future research opportunities. *Int. J. Inf. Manag.* 2020, 52, 10206.

Xiong, H.; Chen, M.; Wu, C.; Zhao, Y.; Yi, W. Research on progress of blockchain consensus algorithm: A review on recent progress of blockchain consensus algorithms. *Future Internet.* 2022, 14, 47.

Xie, S.; Zheng, Z.; Chen, W.; Wu, J.; Dai, H.N.; Imran, M. Blockchain for cloud exchange: A survey. *Comput. Electr. Eng.* 2020, 81, 106–526.

Yli-Huumo, J.; Ko, D.; Choi, S.; Park, S.; Smolander, K. Where is current research on Blockchain technology?—A systematic review. *PLoS One.* 2016, 11(10), e0163477.

Yu, Y.; Liu, S.; Yeoh, P.L.; Vucetic, B.; Li, Y. LayerChain: A hierarchical edge-cloud blockchain for large-scale low-delay industrial Internet of things applications. *IEEE Trans. Ind. Inform.* 2020, 17, 5077–5086.

Zheng, Z.; Xie, S.; Dai, H.N.; Chen, X.; Wang, H. Blockchain challenges and opportunities: A survey. *Int. J. Web Grid Serv.* 2018, 14, 352–375.

Zhang, X.; Xue, M.; Miao, X. A Consensus algorithm based on risk assessment model for permissioned blockchain. *Wirel. Commun. Mob. Comput.* 2022, 2022, 8698009.

3 Exploring Cryptocurrencies – Heading towards the Peak or Valley

Baisakhi Mukherjee and Biswarup Chatterjee

3.1 PREVIEW

With the advent of the digital era, cryptocurrencies are an emerging concept. Based on the academic resources available, it is evident that cryptocurrency is gaining acceptance and is becoming popular in the growth and development process of the economy. Cryptocurrencies are expected to play a pivotal role in the shift from risk management to achieving financial stability, simultaneously expanding opportunities for financial inclusion and fostering innovation in the financial sector (Kelly, 2014). At the time when this chapter was being written, Bitcoin was rising to the level of $26,500, and there was speculation that it could trigger negative market sentiment if it crossed this threshold. Experts have opined that, from the overall technical perspective, the position of Bitcoin is trending slightly towards positivity. This chapter intends to explore the crypto industry and attempts to forecast the position of Bitcoin in the near future.

The World Economic Forum (WEF) reported in July 2022 that cryptocurrencies are a type of virtual asset, primarily composed of cryptographic algorithms and distributed ledger technologies. This chapter intends to deal with virtual assets and online infrastructure such as Bitcoin, which are freely accessed, exposed, and permissionless in nature. The operation of Bitcoin largely depends on a global network of computers to validate transactions and ensure network integrity (Saito, 2016). As an irony of life, cryptocurrencies are misunderstood many times as currencies, but they are not currencies at all (Kelly, 2014). They are conjectural assets whose estimates vary over time. This feature of crypto makes its acceptance and usage more complicated among users. The awareness and regulation of crypto are at its nascent stages, which attracts the academic researchers to dig deeper and develop insights that will prove fruitful for individuals in every industry (Team, 2016).

According to experts in economics, the field of cryptocurrency is not yet researched. There is very little knowledge about the concept, operations, and regulatory policies

DOI: 10.1201/9781003453109-3

of the crypto industry. It has been claimed that the economic impact of cryptocurrencies is still an untouched area (Team, 2016). There are many unanswered questions about the relations between cryptocurrencies and the economy. It is still an open area for the government of countries to decide whether to make it a legal tender for the smooth flow of it in the economy, ban it completely, or develop regulatory policies and frameworks to safeguard the financial stability and growth of investors (World Bank Group, 2018).

This chapter intends to understand the features of cryptocurrencies (BTC), develop a linear regression model to analyse the relationship between the closing price and the volume of BTC transacted, and then proceed with forecasting the average closing price of BTC in the near future. This statistical outcome will help to develop an insight about the position of BTC Patterson, J. (2015).

3.2   LITERATURE REVIEW

Cryptocurrencies represent a decentralized, virtual, and independent monetary system. They liberate money from being reliant on any single currency for transactions, investments, and wealth preservation, thereby breaking free from monopolies (Saito, 2016). They exist beyond the control and regulation of sovereign governments or government entities. These digital currencies are safeguarded by cryptographic systems and ensure a safe and protected internet experience irrespective of monitoring by intermediaries or regulators, such as the Reserve Bank of India (World Bank Group, 2018). Cryptocurrencies are essentially private digital currencies that lack official backing from the Indian government. India's government is in the process of developing its own native online currencies and does not support any of the current cryptocurrencies (Team, 2016). Arguably, this concept was first developed by Satoshi Nakamoto, the creator of Bitcoin, the most popular and widely used of all the cryptocurrencies.

3.3   DIFFERENT CRYPTOCURRENCIES

Apart from Bitcoin, some other popular cryptocurrencies are Ethereum (ETH), Litecoin (LTC), Chainlink (LINK), Cosmos (ATOM), Monero (XMR), Tether (USDT), XRP, Binance Coin (BNB), USD Coin (USDC), Cardano (ADA), Solana (SOL), Dogecoin (DOGE), Tron (TRX), Polygon (MATIC), etc. Many of today's cryptocurrencies have their roots in Bitcoin to some extent. Despite the existence of numerous Bitcoin competitors, the original cryptocurrency still holds its dominant position in terms of adoption and economic worth. To date, none of them has been able to surpass its market capitalization and value (Team, 2016). Since its introduction in 2009, a host of imitators, alternatives, and new technologies based on its blockchain have emerged. These alternative coins are collectively called altcoins. Their purposes have wide ranges – from being a joke/meme to a coin that pays for transactions on a distributed and global virtual machine (Nakamoto. 2008).

3.4   BLOCKCHAIN

The unique technology on which cryptocurrencies work is called blockchain. Unlike other types of databases, a blockchain records transaction data on a publicly distributed ledger (Pilkington, 2016). This data is organized into blocks, which are connected in a sequential chain. Each block also includes information about the previous block, creating a virtually unalterable and easily verifiable chain of time-stamped events (Tapscott & Tapscott, 2016). The blockchain is decentralized and distributed across numerous computers worldwide, making it extremely challenging to tamper with the data stored in the ledger. Since the data on the blockchain isn't held centrally, the failure of one or multiple network participants does not have a catastrophic impact on the entire network (Patterson, 2015). As new blocks are added, they become increasingly difficult to modify, which is a key economic security feature that sets Bitcoin apart (Tapscott & Tapscott, 2016).

3.4.1   Valuation of Cryptocurrencies

Numerous factors exert influence on the value of Bitcoin and other cryptocurrencies. Some of these factors contribute to the significant volatility in their prices. It's very common for their prices to vary by as much as 5% or even 10% in a single day, with smaller cryptocurrencies experiencing even more substantial price swings (Prahalad & Hammond, 2002).

Unlike fiat currencies, cryptocurrencies typically lack the backing of any central authority. Government support plays a crucial role in instilling confidence in the value of a currency among the general public, creating a substantial user base for that currency (Pilkington, 2016). Cryptocurrencies are generally decentralized and derive their value from various sources, including:

- Market forces.
- Production expenses.
- Market liquidity.
- Rivalry.
- Management.
- Compliance.

1. **Market Forces**

The value of cryptocurrencies, like other commodities in the market, is determined by the forces of supply and demand. When the demand for a cryptocurrency exceeds its supply, its price tends to rise, and conversely, if the supply outpaces demand, the price tends to fall (Prahalad & Hammond, 2002).

Cryptocurrencies have well-defined supply mechanisms, with each cryptocurrency disclosing its token creation and destruction plans. For instance, Bitcoin has a fixed maximum supply, capped at 21 million Bitcoins, while Ethereum does not impose such a limit. Some cryptocurrencies employ "burning" mechanisms, where existing tokens are permanently removed from circulation to prevent an excessively large circulating supply and curb inflation. Burning involves sending tokens to an irretrievable address on the blockchain (Stefan Hubrich, 2017). Each cryptocurrency follows its unique monetary policy.

3.5   PRODUCTION EXPENSES

Mining is a process through which new cryptocurrency tokens are produced. The process uses a computer to verify the next block in a blockchain. The persons who do this work are called miners. They use a decentralized network for this. As a form of compensation, miners receive newly created cryptocurrency tokens, along with any transaction fees paid by the parties involved in the cryptocurrency transactions (Pilkington, 2016). This whole process requires computing power for which the miners need to invest in expensive equipment and electricity. The miners need to solve complex mathematical problems in order to verify a block (Kar, 2016). The competition is higher for mining popular cryptocurrencies such as Bitcoin and Ethereum. As a result, mining costs also increase. As the expenses associated with cryptocurrency mining rise, it becomes necessary for the cryptocurrency's value to increase as well (Hopp et al., 2017). As long as the value of the crypto is more than the costs, the miners do not mind and the demand for that particular crypto will be more. As a result, its value will also start increasing (Hanna Halaburda & Guillaume Haeringer, 2018). But once the cost of mining becomes more than the value of the crypto, the miners face a loss and are not interested in mining that particular crypto anymore. Hence, the demand for that crypto goes down. As a result, its value also goes down (Kar, 2016).

3.6   CRYPTOCURRENCY EXCHANGES

Just like any other securities, cryptos also have their own exchanges where they are traded. Prominent cryptocurrencies like Bitcoin and Ether are actively traded on numerous exchanges. It is obvious that cryptocurrency exchanges list the most popular ones (Hoppp et al., 2017).

But some smaller, not-so-popular cryptos are available only on select exchanges. It makes their access limited for the investors interested in them. These exchanges charge fees for the listings of the cryptos (Hoppp et al., 2017).

When a cryptocurrency is listed on multiple exchanges, it increases its demand among investors who are willing to buy it, thus increasing the overall demand. As demand increases, the price goes up (Lagarde Christine (Ed.) 2018).

3.7   COMPETITION

As cryptos have become an alternate investment option for investors, thousands of different cryptocurrencies have entered the market. The entry barrier is relatively low for new competitors (Aisen & Veiga, 2006). But creating a viable cryptocurrency is not an easy job. The success of a cryptocurrency relies on establishing a user network, often achieved by developing a valuable application on the blockchain. When a new entrant in the market gains momentum, it begins to extract worth from the established competition. As a result, the new competitor's token price moves higher (Hoppp et al., 2017).

3.8   INTERNAL GOVERNANCE

Cryptocurrency networks hardly have any set of rules to govern them, and investors generally like stable governance. They feel secure if it is there. In stable governance, rules are relatively hard to change, which provides more stable pricing. Governance can manifest in various ways, including decisions regarding token mining and utilization. But for that purpose, consensus among the stakeholders is very much needed (Aisen & Veiga, 2006).

Cryptocurrency developers often adjust their projects based on the needs and preferences of the user community. However, if the process of updating software to improve protocols is slow, it affects the upside movement of cryptocurrency values. Sometimes, this updating lags for more than 30 days to get implemented. This negatively affects the feelings or attitudes of the present stakeholders (Schuh, 2017).

3.9   REGULATIONS AND LEGAL REQUIREMENTS

When it comes to the legality and regulation of cryptocurrency exchanges, the governments of different countries are a little bit confused. As in the USA, the confusion/dispute is between two government agencies – the Securities Exchange Commission (SEC) and the Commodity Futures Trading Commission (CFTC). The SEC (US Securities and Exchange Commission) considers cryptocurrencies to be securities, similar to stocks and bonds, subject to regulatory oversight. In contrast, the CFTC (US Commodity Futures Trading Commission) categorizes cryptocurrencies as commodities, akin to assets like gold, silver, and petroleum, within their regulatory framework. This difference in classification can lead to varying regulatory approaches and implications for the cryptocurrency market.

Greater clarity in this regard would improve the investors' confidence in cryptos and their values. It would also help in trading other crypto-related financial products in the market (Ammous, 2015; Hoppp et al., 2017).

Regulations are also required for ease of trading cryptocurrency-related instruments such as future contracts, options, etc. It would help investors take short positions. This would also help reduce volatility in crypto prices (Hoppp et al., 2017).

Again, the flip side of too much regulation is that it may have negative impact on the demand for cryptocurrencies. As a result, the prices of cryptos could take a nose dive. But the governments of most countries either do not have clear-cut regulations on cryptos or have not declared trading and investments in cryptos legal for their citizens (Schuh, 2017). All these factors have made the crypto market and prices very volatile at present. Hence, investments in them have become very risky propositions for investors (Ammous, 2015).

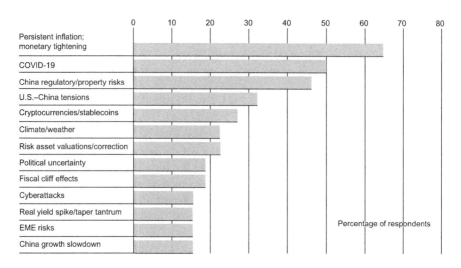

FIGURE 3.1   Autumn 2021: Most-cited potential risks over the next 12 to 18 months. Survey of the US Federal Reserve's May 2022 on risks to financial stability.   Source: United States Federal Reserve Bank of New York.

As per the survey conducted by the US Federal Reserve in May 2022, cryptocurrencies were fifth in the rank, in the most-cited risks to the financial stability at the global level. Figure 3.1 shows the various sources of risk to the financial stability of the world economy.

As per the experts, it has been stated that if crypto currencies are used for any transactions, they will become available and circulate in the economy without any clear regulatory policies, and widespread adoption of the instrument may increase financial stability risks (Schuh, 2017). Due to the absence of monetary policies and controlling authority, such as a central bank for other regulated currencies, cryptocurrencies are not able to utilize the adjustable benefits of monetary policies (Lagarde Christine (Ed.) 2018). Despite the lacunas the crypto assets are evolving rapidly, which could pose a threat to financial stability (O'Sullivan & Sheffrin, 2003). According to the report of the International Monetary Fund (IMF), it has been noted that a strong correlation between Bitcoin and US stocks was observed during the Covid-19 era with the S&P 500 index (O'Sullivan, & Sheffrin, 2003). It has been explored from previous literary works that cryptocurrency is more favoured as a means of investment rather than a means of payment. For instance, since Bitcoin has a limited supply in the market, the price of Bitcoin increases with increasing demand for secured transactions (Hanna Halaburda & Guillaume Haeringer, 2018).

Literature review suggests that 98% of the illegal activities involving financial transactions such as terrorism, ransomware, gambling, etc., are conducted through cryptocurrencies, which is a matter of great concern for every state government. It has been found that crypto is being used in illegal drug trade, hacks, thefts, illegal pornography, and murder for hire. It is suspected that crypto is extensively used in funding terrorist activities, money laundering, and avoids capital controls (Hanna Halaburda & Guillaume Haeringer, 2018).

3.10 INDIAN SCENARIO

The government of India has still not declared cryptos to be legal in India. Hence, no regulations have been laid down so far. Cryptocurrency in India operates without regulation or issuance by any central authority. The absence of specific guidelines for resolving disputes or disagreements when dealing with cryptocurrency is a notable aspect of this unregulated landscape. Hence, if interested persons want to deal in it, they will have to do it at their own risk. But from the statements issued and stances taken by the Governor of the Reserve Bank of India and the Honourable Finance Minister of India, Ms. Nirmala Sitharaman, so far, we can safely say that it is not banned in India till date. The Finance Minister of India has implemented taxes on cryptocurrencies, leading to discussions and debates regarding the legal status of cryptocurrency within the country (Hanna Halaburda & Guillaume Haeringer, 2018). Concurrently, the government has also undertaken efforts to launch India's own cryptocurrency and establish an exchange to oversee the regulation of cryptocurrency trading and investments (Saito, 2014). Until 2022, cryptocurrency operated in a largely unregulated environment. However, the situation altered when the authority introduced a 30% tax on cryptocurrency profits and a 1% tax deducted at the source in the Union Budget of 2022. This clarified the motives of the officials in India. Most of the participants voted in favour of this action and believe that it is the initial step towards achieving official recognition for cryptocurrency within the country (Saito, 2014).

3.10.1 BITCOIN

DeVries (2016) stated that Bitcoin is the most well known and popular cryptocurrency being used. The author also contends that a limited number of Bitcoins are generated, which prevents an excessive supply of the cryptocurrency in the economy and preserves its rarity. Kelly (2014) further explained that the value of Bitcoin is subjective in nature. It means that the value of Bitcoin will remain and continue to increase as long as users have faith or trust in its usage while making payments or other transactions. It has been further acclaimed that Bitcoin does not have any inherent values like gold, diamond, or any other physical objects.

Though cryptocurrencies are at a nascent stage, if accepted they will become the most used and traded currencies in the global economy. As per Team 2016, BitPay is the largest Bitcoin processor in the world and has registered a growth in the transaction rate of 110%. This increase in transaction indicates the growth in the acceptance of the currency. This chapter will explore the position of Bitcoin by using the historical price dataset of BTC-USD from the period 2016–2023. The adoption of cryptocurrencies will attract maximum attention from academic researchers. The factors influencing the awareness and acceptance of cryptocurrencies are yet to be discovered and studied and have a tremendous scope for research (Rothstein & Uslaner, 2005).

According to previous studies, it has been found that Bitcoin is strong by its design of limited availability. It will have diminishing returns every four years until it reaches its maximum number of availability that is 21 million (King 2013). The rarity of Bitcoin is a vital aspect that establishes its worth in the market. The author also claims that the rarity of Bitcoin protects it from an overabundance of the currency in the market, impacting the financial stability and risk (Magro, 2016). Margo also states that cryptocurrencies other than Bitcoin are protected from inflation caused by alterations in government policies, creating a safe cocoon for investors to invest their money. Like other currencies, Bitcoin also suffers from price fluctuations caused by spurious external factors. Both the advantages of being a safe option for transactions and the price instability caused Bitcoin to become the best currency of 2015 using the US Dollar Index, which can be considered a landmark event in the global economy.

It is evident in the literature that the global economy has seen an enormous increase in the volume of Bitcoin transactions. South America has witnessed an increase of 510% in the volume of Bitcoin from the period of 2014–2015.

Bitcoin faces several significant challenges to achieve widespread user acceptance. One major issue is the frequent and substantial fluctuations in its value, which create uncertainty among both users and investors. These value swings erode trust in the cryptocurrency's ability to maintain its worth on a daily basis, undermining overall confidence in these digital currencies. According to a PwC survey conducted in 2015, a significant 83% of respondents had limited to no familiarity with Bitcoin. The absence of centralized control over cryptocurrencies poses a unique challenge. Attempts to address this lack of awareness through advertising campaigns could inadvertently benefit competing companies, making it an unfavourable strategy for marketing plans (McMillan, 2014). Additionally, the cryptocurrency space has witnessed instances of fraud and theft, often stemming from vulnerabilities in exchange company systems. These security breaches frequently make headlines, further discouraging the general public from considering these platforms as safe places to invest their money (McMillan, 2014).

Furthermore, there is a substantial regulatory gap surrounding the use of cryptocurrencies. As long as digital currencies operate in a legal grey area, user acceptance will remain restricted. Users must have confidence that transactions involving cryptocurrencies are legally binding and conform to established laws and regulations. Both markets and governments tend to adapt slowly to emerging technologies, exacerbating the challenges associated with building trust in Bitcoin and other cryptocurrencies (Saito, M. 2016). Another significant challenge facing cryptocurrencies is the complex web of regulations in the United States that must be navigated before achieving widespread user acceptance. The US government has not yet provided a definitive classification for Bitcoin, posing a challenge for many market participants interested in accepting online currencies in business models (PwC, 2015). Virtual assets could potentially fall under different categories, and each of these categorizations can have exceptional consequences for the adoption of Bitcoin (Rothstein & Uslaner, 2005). The global perception of Bitcoin varies from one country to another, with generally positive views influenced by transaction data from Bitpay. In Europe, Bitcoin transactions have surged to a maximum record of 102,221 per quarter (Patterson, 2015), which likely led to the introduction of regulatory measures concerning Bitcoin and other cryptocurrencies. The European Court of Justice has ruled in favour of exempting Bitcoin transactions from value-added tax, essentially acknowledging it as a legal payment method in Europe (Perez, 2015). This indicates that Bitcoin transactions are not subject to taxation in Europe. While this is encouraging for European Bitcoin users, many other major markets still lack essential legislation regarding Bitcoin taxation. The regulatory landscape in the United States, in particular, could potentially have adverse effects on the way Bitcoin transactions are conducted, posing a significant challenge to its recognition as a legitimate currency (McMillan, 2014). There are many grey areas in Bitcoin, which makes it vulnerable and risky in terms of financial stability. Because Bitcoin operates on a public ledger, also known as a blockchain, every transaction is visible to all users who have used Bitcoin. This transparency can make it vulnerable to attacks or hacks due to the ease of access (King, 2013).

The literature has identified a significant security flaw in the design of cryptocurrencies that enables hackers to siphon Bitcoin from exchanges. This caused a serious drop in the volume of Bitcoin. Bitcoin holders sold their Bitcoin in anticipation of theft (Schuh, 2017). These security issues pose a huge hurdle in the process of the acceptance and adoption of Bitcoin in the financial transactions. This chapter intends to regress the closing price of Bitcoin with the volume traded in Bitcoin by developing a regression model.

3.11 METHODOLOGY

To establish a relationship between the closing price and the volume of transactions, simple linear regression OLS model has been developed. As per the literature, the linear regression model relies on five fundamental assumptions. However, in practical applications with our dataset, it is probable that these assumptions may not hold entirely.

> **Linearity**: The expected value of the dependent variable (y) is a linear function of each independent variable when holding the other independent variables constant. To better meet this assumption, we enhance our model by incorporating nonlinear transformations of the features.
> **Independence**: The residuals produced by the fitted model are independent of one another.
> **Homoscedasticity**: The variance of the residuals remains constant in relation to the dependent variable.

Normality: The errors originate from a normal distribution with an unknown mean and variance, which can be estimated from the data. This assumption is employed for calculating "confidence" intervals using well-established analytical expressions.

Multicollinearity: In the case of multiple linear regression, it is crucial for this assumption to hold, meaning there should be minimal or no linear dependence among the predictor variables.

The equation of linear regression can be expressed as,

$$Y = \beta0 + \beta1 X$$

where $Y = Price$ of Bitcoin.
X = Volume of bitcoin transacted.
$\beta0$ = Intercept of the best fit line of regression.
$\beta1$ = Coefficient of the independent variable.

and the assumptions of the models are validated as shown in Figure 3.2.

Proceeding with the time series forecasting procedure has been implemented to fit model into vector using daily seasonality. Time series forecasting is a popular research technique widely used in different areas, especially in high-frequency trading. The time series forecasting model provides a useful edge to the trader by developing an accurate forecasting model that can forecast the future prices of the trading instrument with maximum accuracy. This chapter will focus on developing

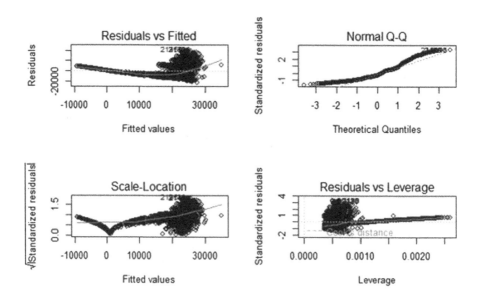

FIGURE 3.2   Validating the assumptions of the model.

an innovative forecasting model using the daily seasonality of the Bitcoin time series data, creating a future data frame for one year and forecasting the future closing price range with the help of the forecasting model.

3.11.1 Data Analysis

For the data analysis, the historical price dataset of Bitcoin for the period January 2016 to September 2023 has been collected from Yahoo Finance. The data analysis has been done using the open software R.

3.11.2 Exploratory Data Analysis

This section on data analysis represents the pattern found in the data retrieved through the visual tools.

Figure 3.3 supports the price volatility of Bitcoin. As presented in the literature review, Bitcoin was the most popular and highly traded cryptocurrency during the period of the pandemic. It is clearly visible in the figure that there is a steep rise in prices during the period 2020–2022.

Figure 3.4 supports the fact that Bitcoin is the most popular among other cryptocurrencies and was popular because of its rarity and price volatility. As presented earlier, Bitcoin was the most popular and highly traded cryptocurrency during the period of the pandemic. It is clearly visible in the figure that there is a steep rise in prices during the period of 2020–2022.

The data is cleaned by removing the missing values and checking the presence of outliers.

Regression model: Linear regression model developed by regressing the closing prices with the volume of transactions of BTC-USD for the period of 2016–2023 (Table 3.1).

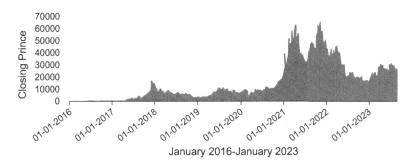

FIGURE 3.3   Pattern of the closing prices of Bitcoin over the period of 2016–2023. Source: Authors.

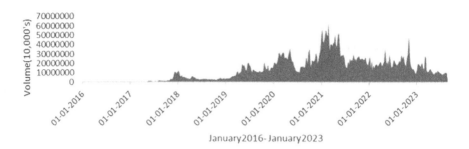

FIGURE 3.4 Pattern of the volume (10,000s) of Bitcoin over the period of 2016–2023. Source: Authors.

TABLE 3.1
2003

Call:
lm(formula = Close ~ lnVolume, data = my_data)

Residuals:

Min	1Q	Median	3Q	Max
−22069	−9714	−3590	6719	42915

Coefficients:

	Estimate	Std. Error	t value	Pr(>|t|)	
(Intercept)	−89845.0	2512.1	−35.77	<2e−16	***
ln Volume	4693.6	110.7	42.39	<2e−16	***

Signif. codes: 0 '***' 0.001 '**' 0.01 '*' 0.05 '.' 0.1 ' ' 1

Residual standard error: 12650 on 2799 degrees of freedom
Multiple R-squared: 0.3909,   Adjusted R-squared: 0.3907
F-statistic: 1797 on 1 and 2799 DF, p-value: <2.2e − 16

H_0: Y_{cp} has no significant relationship with X_{vol}.
H_a: Y_{cp} has a significant relationship with X_{vol}.

3.11.3   REGRESSION FINAL MODEL

$$Y_{cp} = \beta_0 + \beta_1 \log X_{vol}$$

$$Y_{cp} = -89845.0 + 4693.6 \log X_{vol}$$

From the above residual, the *p*-value is $<2.2e - 16$, which does not support the null hypothesis. It can be further interpreted that there is a significant relationship between the price and the volume of transactions of Bitcoin.

3.11.4 FORECASTING PROCEDURE

The forecasting model has been developed in R software by using readxl, stringr, dplyr, and prophet packages. The dataset is cleaned and preprocessed by replacing the null values with 0, and renaming the date and close columns as ds and y, respectively. Model 1 is constructed by utilizing the Prophet library's function "*f(x)*" to fit the model to a vector with daily seasonality.

Model1<-prophet (my_data2, daily.seasonality = TRUE)

To forecast the closing price of BTC-USD, another dataset is created for a time period of 1 year.

Future1<-make_future_dataframe (Model1, periods = (365*1))

The future dataset is tested and trained by Model 1 to forecast. Another vector is created that includes the

Date "Ds",
Standard predicted close price ŷ,
lower estimated close price ŷ_lower,
and upper estimated close price ŷ_upper.
Forecast1<-predict(Model1,Future1)

The predicted price trend is depicted by the slender line, while the dashed black line illustrates the historical price evolution over time in Figure 3.5.

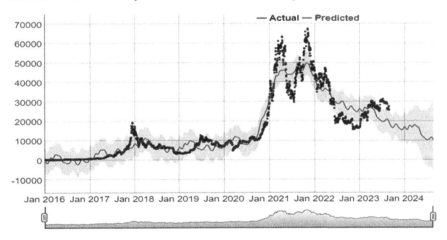

FIGURE 3.5 Plot model estimate. Source: Authors.

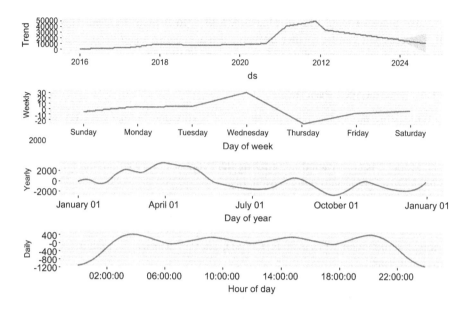

FIGURE 3.6 Components of the forecast 1. Source: Author Generated

Figure 3.6 discusses the separate forecasted elements of the data. These separate forecasted elements represent the seasonal patterns observed on a daily, weekly, and annual basis.

Taking it further, a subset of the data frame was created for half of each month of the year, and a table of prices for 1 year table is created from the trained dataset. For a detailed analysis, the new data frame was filtered between the time range from 15th May 2022 to 15th June 2023, as the recent trend will provide a better outlook.

3.11.5 TRUNCATED DATA FRAME

	ds	yhat	yhat_lower	yhat_upper
2327	2022-05-15	33146.15	26882.8485	39561.72
2358	2022-06-15	30267.90	24546.8472	36723.84
2388	2022-07-15	29026.95	21749.9966	35101.57
2419	2022-08-15	30305.35	24136.0582	36700.56
2450	2022-09-15	26811.49	20421.2242	32648.97
2480	2022-10-15	27008.18	20628.1034	33434.92
2511	2022-11-15	26276.44	19985.1692	32701.60
2541	2022-12-15	24792.60	18733.4692	31432.77
2572	2023-01-15	25856.13	19561.2186	32147.43
2603	2023-02-15	26985.40	20880.2528	32852.36
2631	2023-03-15	26464.96	20111.2079	32404.81
2662	2023-04-15	26713.07	20418.1063	33058.85

```
|2692 |2023-05-15 | 23905.10| 17664.9025|   30390.72|
|2723 |2023-06-15 | 20927.05| 14350.8622|   27359.97|
|2753 |2023-07-15 | 19730.97| 13218.0404|   26082.63|
|2784 |2023-08-15 | 21009.81| 14739.8575|   27466.77|
|2815 |2023-09-15 | 17562.34| 11443.6623|   23411.37|
|2845 |2023-10-15 | 17681.69| 11828.2918|   23602.52|
|2876 |2023-11-15 | 17028.94| 10517.4298|   23738.87|
|2906 |2023-12-15 | 15513.70| 8635.3736|    22555.21|
|2937 |2024-01-15 | 16593.45| 9132.8090|    23817.97|
|2968 |2024-02-15 | 17610.36| 10362.2174|   25995.06|
|2997 |2024-03-15 | 17186.85| 8342.5052|    26929.78|
|3028 |2024-04-15 | 17383.39| 7050.0521|    27806.74|
|3058 |2024-05-15 | 14495.13| 3008.4600|    26355.11|
|3089 |2024-06-15 | 11608.73| -855.6246|    25157.73|
|3119 |2024-07-15 | 10447.04| -3951.0608|   26114.10|
|3150 |2024-08-15 | 11675.33| -5136.3941|   28914.79|
```

Mean of *truncated ŷ*

```
|          x|
|--------:|
| 21571.95|
```

The standard forecasted average closing price for the next year(2024) is forecasted as $21571.95.

Mean *truncated ŷ* $_{Lower}$

```
|          x|
|--------:|
| 13871.64|
```

The lower bound forecasted average closing price of BTC-USD for the next year (2024) is $13871.64.

Mean *truncated ŷ* $_{upper}$

```
|          x|
|--------:|
| 29445.31|
```

The upper bound forecasted average closing price of BTC-USD for the next year (2024) is $29445.31.

The projected outcomes suggest a reasonable range of expected Bitcoin closing prices for the upcoming year. Nonetheless, it's important to note that despite the package's title, this analysis doesn't provide a foolproof, 100% accurate glimpse into the future and should be approached with caution due to the inherent volatility of the Bitcoin market. Additionally, unforeseen and rare events, often referred to as black swan events, should always be taken into consideration.

Regarding Bitcoin's role as an asset, this regression model takes into account certain factors such as historical retail and institutional buying patterns, seasonal variations, and holidays, as well as network difficulty (which reflects the level of challenge in mining Bitcoin blocks and shows a somewhat staggered connection to Bitcoin's price).

Nonetheless, it's important to note that this forecasting model may not comprehensively account for the effects of shifts in the Federal Funds Rate. Employing multiple linear regression analysis could be a more appropriate method for accurately assessing the impact of Federal Funds Rate fluctuations on the value of Bitcoin.

3.12   ADVANTAGES OF CRYPTOCURRENCY

As it is a technology-based financial product, cryptocurrencies have gained popularity among investors worldwide in recent years, thanks to the massive technological development in software, hardware, and information technology. Cryptocurrencies facilitate the transfer of funds without the need for banks and other financial institutions, making it particularly convenient for international and cross-border transactions. Consequently, Bitcoin and Ethereum have gained a favourable position in this regard. The advantages of cryptocurrencies can be summarized as follows:

1. Protection against Inflation

Inflation reduces the actual values of products as well as currencies. But investors consider cryptocurrency as a product that gives protection against inflation. Most cryptocurrencies use the same mechanism to maintain immunity against inflation (*B Scott - 2016 - econstor. eu*)

2. High Speed of Transactions

Electronic money transfers typically take several hours to even a few days, depending on various factors such as the geographical distance between the sender and recipient, the location of the origin and destination, especially for international and cross-border transactions, and other variables. In contrast, cryptocurrency transactions can be conducted within a matter of minutes, providing a hassle-free and appealing option to many.

3. Low Cost of Transactions

Cryptocurrencies do not need the intervention of any third parties, agencies, or payment gateways like VISA to complete transactions. Hence, they can help transfer funds globally almost free of cost. Transaction costs associated with cryptocurrency can be minimal or even negligible (Investopedia: Adam Hayes, updated April 23, 2023).

4. **Decentralized Model of Network**

Cryptocurrencies leverage blockchain technology, which represents a novel decentralized monetary model. They contribute to breaking the monopoly of any specific currency and liberate money from the control of central authorities. As no government or government organizations can control their flow, investors find cryptocurrencies to be a secure and safe investment option and transaction product.

5. **Diversified Portfolio**

For any investment portfolio, the main "mantra" is diversity – do not put all your eggs in one basket. Hence, rational investors always look for newer, safer, and more profitable investment options. Investments in cryptocurrency can generate profits as it is becoming very popular and can give the required diversity in an investment portfolio. But the main problem with investment in cryptos is its price volatility. As the prices of cryptos have less correlation with other market forces like the prices of shares and bonds, it makes cryptos a good investment option for portfolio diversification. This low-price correlation may give a handsome return to crypto investments(Coursera, updated on June 16, 2023).

6. **Easy Accessibility**

Unlike the traditional investment options where the investors need to provide lots of personal information or data (KYC), sometimes it becomes tiring, cumbersome, irritating, and risky for the investors to divulge or share so much personal data or information with the authorities in light of the many fraudulent activities occurring nowadays. Investing in cryptocurrencies requires investors to have only a computer or smartphone with an internet connection, making it a hassle-free, streamlined, secure, and time-efficient process (Coursera, updated on June 16, 2023).

7. **Safety and Security**

In a cryptocurrency blockchain system, there is a private key for every crypto wallet. This key is nothing but a set of codes. No crypto wallet can be accessed without this key. This gives cryptos the required security and safety. If an investor forgets or loses this key, then he cannot operate/recover his funds.

Another benefit of cryptocurrencies is that transactions are secured by the blockchain system, which is a decentralized network of computers that validate and verify these transactions. Hence, if any problem or hacking arises in one terminal, it cannot affect the whole network of the blockchain system. As a result, the funds remain secured. The security of the investors can be enhanced if they keep their crypto assets in their own wallets (Coursera, updated on June 16, 2023).

8. **Operational Transparency**

Blockchain form of operations provide transparency in cryptocurrencies. This transparency makes the whole operational system simple and boosts investors' confidence as it makes it free of any scope of corruption. Investors can easily monitor their money transfer transactions by using a blockchain explorer on the platform, allowing them to track live transfers in real time (Coursera, updated on June 16, 2023).

9. **Privacy**

As there is no intervention from any government or agency, the whole system has a certain level of privacy. The investors can transact using pseudonyms with the help of their wallet keys or addresses, i.e. without divulging their original identities (Coursera, updated on June 16, 2023).

10. **Conversion vis-à-vis Other Currencies/Liquidity**

Cryptocurrencies can be converted into other widely used and popular currencies such as USD, Euro, Yen, Pound, etc., through the cryptocurrency exchanges. This gives immense liquidity to the investors for their funds (Coursera, updated on June 16, 2023).

3.13   DISADVANTAGES/DRAWBACKS OF CRYPTOCURRENCY

1. Though cryptocurrencies are claimed to be pseudonymous, the government or government agencies can track the transaction trails if required and hence the possibility of intervention.
2. Many blockchains operate on the proof-of-work consensus mechanism. To participate in such networks, individuals are often obligated to utilize powerful ASIC computers that consume significant amounts of electricity and generate heat, making it a costly endeavour.
3. Most of the cryptocurrencies do not have specific policies regarding transactions, such as no refund or cancellation policies (Published in: 2021 1st International Conference on Computer Science and Artificial Intelligence (ICCSAI), Jakarta, Indonesia).

3.14   CONCLUSION AND FUTURE SCOPE OF RESEARCH

Cryptocurrency appears to have moved beyond the initial phase of adoption typically associated with new technologies, much like motor vehicles did in their early days. Bitcoin has initiated the establishment of its own specialized market, which could either propel virtual assets further into the mainstream or potentially lead to their downfall.

The Bitcoin community is proactively striving to integrate itself into mainstream usage by fostering innovation and addressing longstanding challenges. Numerous alternative cryptocurrencies have emerged, each possessing its unique features but deemed valid in their own regard. The future potentially expects a substantial role for cryptocurrencies as prominent currency solutions, with Bitcoin leading the path for these virtual currencies to grow. Significantly, both the European and Latin American markets are witnessing a surge in Bitcoin transactions, signifying its growing legitimacy (Schuh, 2017).

There are numerous additional topics to explore concerning Bitcoin and cryptocurrencies. In-depth studies should be conducted to investigate the economic consequences of Bitcoin on established fiat currencies, comparing with the outcomes of the countries that are adopting state-sponsored cryptocurrencies. Cryptocurrency's potential for facilitating microtransactions could address economic challenges that traditional state-sponsored currencies might not resolve, although this requires thorough market and economic analysis.

Moreover, blockchain technology, which underlies Bitcoin, holds promise for applications (Hileman, 2016). Smart agreements are controlled transactions triggered by specific conditions; a task traditionally handled by entire company accounting departments. Additionally, there is potential for subsequent evolution.

This emerging frontier is still relatively new and mostly uncharted, with a predominant presence of various media forms. It's possible that other types of digital assets could gain popularity, much like music and cryptocurrencies. Just eight years ago, digital currency was virtually unknown, and the introduction of Bitcoin revolutionized the landscape. Cryptology, the fundamental science underpinning Bitcoin and all virtual assets, has the potential to drive fresh and exciting digital innovations.

REFERENCES

Aggarwal, S., Santosh, M., & Bedi, P. Bitcoin and Portfolio Diversification: Evidence from India 99–115.

Aisen, A., & Veiga, F. J. (2006). Does political instability lead to higher inflation? A panel data analysis. *Journal of Money, Credit and Banking*, 2006, 1379–1389. Retrieved from https://www.jstor.org/stable/3839011.

Anning, F. 2018. An Assessment of the Survival & Regression Analysis of Exchange Closures in Bitcoin Operations. *SSRN Electronic Journal.* https://doi.org/10.2139/ssrn.3197758

Ammous. (2015). Economics beyond financial intermediation: Digital currencies' possibilities for growth, poverty alleviation, and international development. *The Journal of Private Enterprise*, 3.

Ammous. (2018). Can cryptocurrencies fulfil the functions of money? The quarterly review of economics and finance. Advance online publication. https://doi.org/10.1016/j.qref .2018.05.010.

Bank Indonesia. (2018). Bank Indonesia Warns All Parties Not to Sell, Buy, or Trade Virtual Currency. Retrieved from https://www.bi.go.id/en/ruang-media/siaranpers/Pages/sp _200418.aspx.

Barham, V., Boadway, R., Marchand, M., &Pestieau, P. (1995). Education and the poverty trap. *European Economic Review*, 39, 1257–1275. https://doi.org/10.1016 /0014- 2921(94)00040-7.

Beck, T., & Demirguc-Kunt, A. (2006). Small and medium-size enterprises: Access to finance as a growth constraint. *Journal of Banking & Finance*, 30, 2931–2943. https://doi.org /10.1016/j.jbankfin.2006.05.009.

Bearman, J. (2015, May). The Untold Story of Silk Road, Pt. 1. Retrieved from Wired.com Website: https://www.wired.com/2015/04/silk-road-1/.

Benigno, P. (2009). Price stability with imperfect financial integration. *Journal of Money, Credit and Banking*, 41, 121–149. https://doi.org/10.1111/j.1538–4616.2008.00201.x.

Bovaird, C. (2016, June 24). Bitcoin Rollercoaster Rides Brexit as Ether Price Holds Amid DAO Debacle. Retrieved June 2016, from CoinDesk Website: http://www.coindesk .com/bitcoin-brexit-ether-pricerollercoaster/.

Christine, L. (Ed.). (2018). Winds of Change: The Case for New Digital Currency: International Monetray Fund.

Darlington, J. K., III. (2014). *The Future of Bitcoin: Mapping the Global Adoption of World's Largest Cryptocurrency Through Benefit Analysis*. University of Tennessee, Knoxville. Retrieved from https://trace.tennessee.edu/utk_chanhonoproj/1770.

DeVries, P. (2016). An analysis of cryptocurrency, Bitcoin, and the future. *International Journal of Business Management and Commerce, 1*, 1–9.

Dorfleitner, G., Hornuf, L., Schmitt, M., & Weber, M. Definition of FinTech and Description of the FinTech Industry. In FinTech in Germany, pp. 5–10. https://doi.org/10.1007/978 -3-319-54666-7_2.

Fernholz, T. (2015). Terrorism Finance Trackers Worry ISIS Already Using Bitcoin. Retrieved from https://www.defenseone.com/threats/2015/02/terrorism-finance-trackers-worry-isis-already-using-bitcoin/105345/.

FSB. (2022). Assessment of Risks to Financial Stability from Crypto-Assets. Retrieved May 6, 2022, from https://www.fsb.org/2022/02/ assessment-of-risks-to-financial-stabili ty-from-crypto-assets/.

Gorodnichenko, Y., & Schnitzer, M. (2013). Financial constraints and innovation: Why poor countries don't catch up. *Journal of the European Economic Association*, 11, 1115–1152. https://doi.org/10.1111/jeea.12033.

Halaburda, -H., & Haeringer, G. 2018. Bitcoin and blockchain: What we know and what questions are still open. *SSRN Electronic Journal*. https://doi.org/10.2139/ssrn.3274331.

Hofman, A. (2014, March 6). The Dawn of the National Currency – An Exploration of Country-Based Cryptocurrencies. Retrieved from Bitcoin Magazine Website: https:// bitcoinmagazine.com/articles/dawnnational-currency-exploration-country-based -cryptocurrencies-1394146138.

Hileman, G. (2016, January 28). State of Bitcoin and Blockchain 2016: Blockchain Hits Critical Mass. Retrieved from Coindesk Website: http://www.coindesk.com/state-of -bitcoin-blockchain-2016/.

Hubrich, S. 2017. 'Know when to Hodl Em, know when to Fodl Em': An investigation of fac-tor based investing in the cryptocurrency space. *SSRN Electronic Journal*. https://doi .org/10.2139/ssrn.3055498.

Investopedia, "What Is a Permissioned Blockchain?" (2022). https://www.investopedia.com /terms/p/permissionedblockchains.asp#:~:text=A%20permissioned%20blockchain %20is%20a,certificates%20or%20other%20digital%20means. Accessed 11 June 2022.

Jaag, C., & Bach, C. (2015). Cryptocurrencies: New Opportunities for Postal Financial Services. Retrieved from www.swiss-economics.ch.

Jaag, C., & Bach, C. (2015). Cryptocurrencies: New Opportunities for Postal Financial Services. Retrieved from www.swiss-economics.ch.

Johnson, T. Some Implications of a Pragmatic Approach to Finance. In *Ethics in Quantitative Finance*, pp. 271–293. http://dx.doi.org/10.1007/978-3-319-61039-9_13.

Kajtazi, A., & Moro, A. 2018. The Role of Bitcoin in Well Diversified Portfolios: A Comparative Global Study. *SSRN Electronic Journal*. https://doi.org/10.2139/ssrn .3261266.

Kar, I. (2016, June 30). Everything You Need to Know about the Bitcoin „halving" event. Retrieved from Quartz website: http://qz.com/681996/everything-you-need-to-know -about-the-bitcoin-halving-event.

Kelly, B. (2014). *The Bitcoin Big Bang: How Alternative Currencies Are about to Change the World*. Wiley.

King, R. S. (2013, December 17). By Reading This Article, You"re Mining Bitcoins. Retrieved from Quartz.com Website: http://qz.com/154877/by-reading-this-page-you-are-mining -bitcoins/.

Lo, W. (2014). *Bitcoin as Money?* Federal Reserve Bank of Boston.

McMillan, R. (2014, March 3). The Inside Story of Mt. Gox, Bitcoin"s $460 Million Disaster. Retrieved from Wired.com Website: http://www.wired.com/2014/03/bitcoin-exchange/.

Muchlis Gazali, H. M., Hafiz Bin Che Ismail, C. M., & Amboala, T. *Exploring the Intention to Invest in Cryptocurrency: The Case of Bitcoin*, pp. 64–68. https://doi.org/10.1109/ ict4m.2018.00021.

Nakamoto. (2008). *Bitcoin: A Peer-to-Peer Electronic Cash System*.

Olken, B. A. (2006). Corruption and the costs of redistribution: Micro evidence from Indonesia. *Journal of Public Economics*, 90, 853–870. https://doi.org/10.1016/j.jpubeco .2005.05.004.

Nica, Octavian, Piotrowska, Karolina, Schenk-Hoppp, Klaus Reiner. 2017. Cryptocurrencies: Concept and Current Market Structure. *SSRN Electronic Journal*. https://doi.org/10 .2139/ssrn.3059599

O'Sullivan, A., & Sheffrin, S. M. (2003). *Economics: Principles in Action*. Prentice Hall.

Pagano, M. S., & Sedunov, J. 2018. Bitcoin and the Demand for Money: Is Bitcoin More Than Just a Speculative Asset? *SSRN Electronic Journal*. https://doi.org/10.2139/ssrn .3293998.

Patterson, J. (2015, August 4). Bitcoin: A New Global Economy. Retrieved from Bitpay Website: https://blog.bitpay.com/bitcoin-a-new-global-economy/.

Pieters, G., & Vivanco, S. (2017). Financial regulations and price inconsistencies across Bitcoin markets. *Information Economics and Policy*, 39, 1–14.

Pilkington, M. (2016). Blockchain technology: Principles and applications. In F. X. Olleros and M. Zhegu (Eds.), *Research Handbooks in Business and Management Series*, pp. 225–254.

Prahalad, C. K., & Hammond, A. (2002). Serving the world's poor, profitably, 80, 48– 58. Retrieved from https://barnabys.blogs.com/files/serving-the-poor--prahalad -hammond-.pdf.

PwC. (2015, August). Money Is No Object: Understanding the Evolving Cryptocurrency Market. Retrieved from PricewaterhouseCoopers, LLP. Financial Services Website: https://www.pwc.com/us/en/financialservices/publications/assets/pwc-cryptocurrency -evolution.pdf.

Rothstein, B., & Uslaner, E. M. (2005). All for all: Equality, corruption, and social trust. *World Politics*, 58, 41–72. https://doi.org/10.1353/wp.2006.0022.

Rysman, M., & Schuh, S. (2017). New innovations in payments. *Innovation Policy and the Economy*, 17, 27–48.

Saito, M. (2016, February 18). Exclusive: Amazon Ēxpanding Deliveries by Its 'On-demand' Drivers. Retrieved from Reuters Website: http://www.reuters.com/article/us-amazon -com-logistics-flex-idUSKCN0VR00O.

Sutiksno, D. U., Ahmar, A. S., Kurniasih, N., Susanto, E., & Leiwakabessy, A. 2018. Forecasting historical data of Bitcoin using ARIMA and α-Sutte indicator. *Journal of Physics: Conference Series*, 1028, 012194.

Tapscott, D., & Tapscott, A. (2016). *Blockchain Revolution: How the Technology Behind Bitcoin is Changing Money, Business, and the World*. Penguin Random House LLC.

Team, B. (2016, January 20). Understanding Bitcoin's Growth in 2015. Retrieved from Bitpay Website: https://blog.bitpay.com/understanding-bitcoins-growth-in-2015/.

US Securities and Exchange Commission, "Leveraged Investing Strategies – Know the Risks Before Using These Advanced Investment Tools", 10 June 2021: https://www.sec.gov/oiea/investor-alerts-and-bulletins/ib_ leveragedinvesting. Accessed 29 June 2022.

Bracamonte, V., & Okada, H. (2017). An exploratory study on the influence of guidelines on crowdfunding projects in the ethereum blockchain platform. In *Lecture Notes in Computer Science*, pp. 347–354.

Wood, G. (2014). Ethereum: A Secure Decentralised Generalised Transaction Ledger. Retrieved from http://gavwood.com/paper.pdf.

World Bank. (2018). Remittance Prices Worldwide: Making Markets More Transparent. Retrieved from http://remittanceprices.worldbank.org/.

Wilson, T, Howcroft, E., & Lang, H., (2022). "Bitcoin Stabilizes after Heavy Losses But Pessimism Reigns in Crypto Markets. *Reuters*, 14 June 2022. Retrieved from https://www.reuters.com/technology/no-let-up-crypto-slide-celsius-halt-leavesinvestors-pan-icking-2022-06-14/. Accessed 29 June 2022.

World Bank Group. (2018). *Poverty and Shared Prosperity Report: Piecing Together the Poverty Puzzle*. Retrieved from World Bank Group website: https://openknowledge.worldbank.org/bitstream/handle/10986/30418/9781464813306.pdf.

World Economic Forum Global Future Council on Cryptocurrencies, "Cryptocurrencies: A Guide to Getting Started", June 2021. Retrieved from https://www3.weforum.org/docs/WEF_Getting_Started_Cryptocurrency_2021.pdf. Accessed 30 May 2022.

Zhao, Y., & Duncan, B. The impact of crypto-currency risks on the use of blockchain for cloud security and privacy. *International Conference on High Performance Computing & Simulation (HPCS)*, 677–684. https://doi.org/10.1109/hpcs.2018.00111.

4 Blockchain and Its Impact towards Technology Adoption

Arpit A. Jain, Savita Gandhi, and Sarla Achuthan

4.1 INTRODUCTION

Technology has impacted human life in different ways. Blockchain technology is one such technology that has the potential to make human lives better. It has revolutionized the way data are stored. It is a concept and execution method that has sparked a lot of interest across a wide range of companies and sectors. Data are incredibly important in today's environment. Therefore, it is critical to protect private data. Web and mobile applications are the key sources of data collection. By accepting their terms and conditions, users' personal information is sometimes shared with their server. The disadvantage of using any web-based interface is that the data are shared with a server with centralized storage. Now imagine a situation, where data are not stored centrally. In these circumstances, neither a person such as a database administrator nor a machine can tamper with it, hence privacy can never be protected.

Blockchain technology has provided the solution to manage all aspects of privacy and non-tampering of data. Blockchain is a decentralized network of nodes. It works on the concept of distributed ledger allowing data to be kept in a distributed manner across several nodes (Neha et al., 2021). These nodes are linked together to maintain the proper formation of data. It operates on the threefold principle of secure storage, secure transaction verification, and secure chain of blocks. The data are stored in the form of blocks. Each block is made up of one or many transactions. These blocks, linked together using cryptographic hashes, keep a continuous and immutable record of all network transactions.

The cryptographic hashes are algorithm-generated hex-decimal values on encryption of data. Cryptographic hash function is a mathematical function that plays a vital role in the security of data. It takes the message as an input and uses a mathematical formula to generate an encrypted message. The generated string is a combination of letters and numbers. The size of the generated string depends on the algorithm used for the encryption. The notion of hash ensures the security and integrity of the transaction/message in the emerging blockchain technology.

DOI: 10.1201/9781003453109-4

SHA-256 is a hash mathematical function that is used in one of the most famous blockchain technologies, i.e. Bitcoin, developed by the National Security Agency, USA (Neha A. Samsir & Arpit A. Jain, 2022). It is a 256-bit algorithm which is widely used by Bitcoin in blockchain technology. Another mathematical function is Keccak widely adopted by Ethereum. It is developed by a team, of Bertoni, Daemen, and Assche, as an outcome of the NIST hash function competition in 2012. According to the information available on Wikipedia, blockchain can be described as "a distributed ledger with growing lists of (*blocks*) that are securely linked together via cryptographic hashes." The keyword blockchain can be described by using the following definition: "*A blockchain is a chain of transactions mapped sequentially to distribute blocks consisting of information arranged in a secured and non-mutable manner.*"

In Figure 4.1, the concept of blockchain enhances the productivity of companies and is rapidly changing the competitive landscape of the business. Enterprise resource planning integrated with the blockchain mechanism has given potential to business process management and opened more options towards innovative business opportunities. Cryptocurrency is one of the most significant applications of blockchain technology. Virtual currency has been introduced in this planet via blockchain technology. Blockchain technology has been essential to the development and maintenance of cryptocurrencies. From inception to management of cryptocurrency, blockchain technology has played a vital role. The main advantage of cryptocurrencies is that transactions can be authorized and approved without the involvement of third parties or intermediaries like banks or financial corporations (Bansod S & Ragha L 2022). These cryptocurrencies are virtual tokens, and each token's value is based on the volume of buyers and sellers. Blockchain technology, which provides a secure, public, and decentralized way of recording transactions, powers the majority of cryptocurrencies. A few governments throughout the world are skeptical about cryptocurrencies as legal tender, whereas a few of the characteristics of cryptocurrencies that make them more potent include security, decentralization, and accessibility on a global scale.

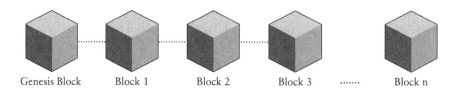

Genesis Block   Block 1   Block 2   Block 3   ....   Block n

FIGURE 4.1   Conceptual view of blockchain. Source: Authors.

4.1.1 HISTORY OF BLOCKCHAIN

Decentralization and cryptography are known concepts for technologists since several years. From the year 1990, people have been using these terminologies according to the requirement and for work efficiency. Though the word "Blockchain" became popular in the market in the year 2008, with Satoshi Nakamoto's conceptualization of the first decentralized blockchain, significant work on blockchain had been done much earlier. For example, way back in 1991, two researchers Stuart Haber and W. Scott Stornetta (Akhtar et al., 2019) outlined an untemperable document time stamp system. Satoshi then showed the power of blockchain to the world by introducing a cryptocurrency named Bitcoin through a white paper (Sturart Haber, Stornetta, & W. Scott 1991). In the early days of Bitcoin, the concept of digital currency had no physical mark value. When Laszlo Hancez purchased two pizzas by paying 10,000 Bitcoin (BTC), the awareness and understanding of the applicability of Bitcoin in the real world increased (Nakamoto, S. 2008). By the late 2013s Bitcoin was a currency that was known to many individuals. People began investing in Bitcoin by understanding the value of digital currency.

Professionals then started studying and researching the technology and analyzing its impact on business applications. By the year 2015, Vitalik Buterin had designed and released Ethereum, a new digital currency. By that point, everyone was fairly familiar with hash code, transaction management, smart contracts, service-level agreements, and block chaining. The need for blockchain technology has since grown, and a wide range of industries have adopted it. As of July 15, 2022, there are 748,655,000 total transactions on the 'Blockchain.com' application (Panda et al., 2021), and that number is expected to rise extremely quickly. Blockchain is not just used in the realm of cryptocurrencies. Data management, enterprise resource planning, and supply chain management are prioritized more. The use of blockchain varies depending on the requirements of the system. This technology's potential will undoubtedly lead to solutions in a variety of fields. The global blockchain market is anticipated to be worth $1,431.54 billion by 2030, increasing at a CAGR of around 85.9% between 2022 and 2030, according to the data published by "BUY BITCOIN WORLDWIDE" (Neha A. Samsir and Arpit A. Jain 2022). Blockchain has the potential to save more than $10 billion annually or more than 30%, on the banking infrastructure expenditures.

4.1.2 NEED FOR BLOCKCHAIN

Blockchain technology has gained significant traction in the market due to several compelling features and benefits. The prominence of blockchain in the market is increasing continuously. The necessity for blockchain in corporate and commercial applications has increased as a result of the many capabilities it offers like decentralization, privacy, security, non-mutability, and many more. The requirement for decentralization in data management necessitates the use of blockchain. It is a form of chain that works on the concept of decentralization. In blockchain many nodes are interconnected in a network and data are consecutively added to such nodes.

Maintaining data in a decentralized form increases transparency and the level of security. This also reduces the risk of a single point of failure. In today's world, when business organizations, governments, and individuals engage in a multitude of online activities, the management and security of these data have become paramount. Thus, security is one of the major requirements in data management for business applications and blockchain meets it by adhering to the idea of non-mutability, which means that the data once written are non-editable. Also, in order to create safe digital environments, it is essential to adhere to the fundamental concepts of authorization and authentication.

Authorization ensures that only authorized users are granted access to specific resources or actions, while authentication verifies the identity of users seeking access. Blockchain also adheres to privacy by following the principles of authorization and authentication. Due to the internet's rapid growth in data, data management, security essentials are one of the major requirements of the industry. The right security essentials will lead towards a more secured environment over the internet. Traditional centralized data management systems frequently have trouble finding a balance between granting access to legitimate parties and protecting data from intrusions, breaches, and manipulations. Blockchain technology can be a game-changing transformative solution for this situation. In essence, blockchain's emphasis on authorization, authentication, decentralization, transparency, and cryptographic security fulfills today's demand for robust data management and security necessities. Blockchain not only improves the security of online interactions by offering a secure and transparent platform for managing transactions, data, and digital identities, but it also paves the way for a more trust-based and efficient digital ecosystem. As enterprises battle with data security concerns, blockchain technology emerges as a possible way to build a more secure environment over the internet.

4.1.3 LITERATURE REVIEW

The blockchain technology, first linked to cryptocurrencies, has seen significant advancements, expanding its scope beyond its initial applications which are associated with digital money transactions. For various domains, it also offers a resilient and decentralized framework that guarantees both security and transparency. The utilization of blockchain technology is considered a fundamental element in the development of many applications. Several scholars (KrstićMarija and Krstić Lazar 2020) have conducted studies on different elements of blockchain technology. Samsir Neha and Jain Arpit (2021) provide a comprehensive overview of blockchain technology and its applications in the healthcare sector. They discuss the fundamental architecture, key features, and various consensus algorithms of blockchain technology. Their study provides a significant contribution to the field by analyzing the blockchain paradigm, emphasizing its potential to enhance system robustness and decentralization. They further discuss the widespread adoption of blockchain in various sectors namely healthcare, education, and finance, focusing mainly on applications of healthcare. They also adhere to the challenges associated with implementing blockchain systems for the healthcare industry. Their study contributes to a better understanding of blockchain's significance and its core concepts for designing the blockchain along with the challenges for the implementation of the blockchain system.

Bansod Smita and Ragha Lata (2022) raise the concern of critical technological requirements for ensuring data privacy and protection in blockchain technology. They suggest security as one of the most important concern for the data. Many online transactions in today's contactless digital world are sensitive and require security, for which blockchain gives promising solution. They also discuss the increasing emphasis on user privacy rights in existing and proposed data protection regulations across various countries. Their study also explores the use of Privacy Enhancing Techniques for cryptocurrencies by using efficient crypto-privacy algorithms. Their study offers significant contributions to the understanding of the dynamic field of data security and privacy in the digital age, serving as a basis for future progress in the realm of blockchain technology.

Mukherjee Pratyusa and Pradhan Chittaranjan (2021) provide a comprehensive analysis of the historical development of blockchain technology. They include detailed explanations of two fundamental concepts in blockchain, namely the consensus mechanism and the structure of a block. They also discuss about the evolutionary phases of blockchain technology with the characteristics of each generation. They define "Blockchain" as secured linked ledger created through blocks of hash functions. They further elaborate upon the fundamental principles behind blockchain technology, including cryptography, decentralization, immutability, and transparency. They then discuss the inclusion of descriptive test scenarios within the realm of supply chain management, with the objective of improving understanding of the practical implementations of blockchain technology in supply chain management.

Kamath Reshma (2018) provides a comprehensive study, which demonstrates the key part that blockchain technology can play in the revolutionary process of improving supply chain management as well as food safety. The author discusses about the adoption of blockchain technology in food supply chain management system as a future for maintaining global food ecosystem. The study emphasizes the significant cost reductions and enhanced visibility in the management of the food supply chain that Walmart effectively accomplished. In the study, the author mainly focuses on challenges as well as opportunities in implementing blockchain technology for the food supply chain management ecosystem on a global scale. She examines IBM's blockchain solution, built upon the Hyperledger Fabric framework citing successful implementation by Walmart of two blockchain pilots' programs in both China and the United States of America.

Samsir N and Jain A (2022) discuss about the working of blockchain by understanding the structure of block and transaction. They present the significance of consensus algorithms as well as the workings of these algorithms in depth. In addition to this, they also present an in-depth review of a few important consensus algorithms, including proof of work (PoW), proof of stake (PoS), practical Byzantine fault tolerance (PBFT), delegated proof of stake (DPoS), and Tendermint. Their study provides insight into the relative merits of different algorithms by analyzing and contrasting them.

4.1.4 CHALLENGES OF BLOCKCHAIN TECHNOLOGY

Technology always helps people to work effectively and efficiently. While there are many advantages to technology, it is vital to also take into account any potential drawbacks, such as privacy problems, digital inequality, and the effect of automation on jobs majorly. The difficulty for society is to balance the benefits of technology with any potential downsides. The challenges are quite often resolved with the numerous benefits of technology [7]. But it is always required to incorporate the challenges for understanding the technology in a better way. Figure 4.2 illustrates the challenges associated with blockchain, which are discussed in greater detail as follows:

i) **Concern of Privacy**: Our lives are now a part of the internet through various digital channels or it can well be said that the internet has become a prominent factor in our lives. Through various digital platforms, people are sharing data continuously. Due to this, security and privacy have become a major concern today. People even do not know where and for what purpose their data are in use. Nowadays, especially after Covid-19, many personal transactions pertaining to finance, commerce, business, and healthcare are all done online leading to increase of data on the internet (Miglietti et al., 2019). As a result, ensuring the privacy and security of sensitive data are a significant concern.

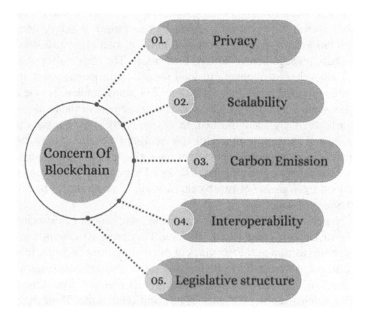

FIGURE 4.2   Concerns of blockchain. Source: Authors.

ii) **Concern of Scalability**: Blockchain networks sometimes struggle to have trouble processing a lot of transactions rapidly. As every single data needs to be mapped in the chain of blocks, the performance of the network may deteriorate as more transactions are added to the blockchain resulting in delay of the mining of blocks. Also adding data/block in a blockchain comes with a mining cost.

iii) **Concern of Carbon Emission**: Various blockchain consensus algorithms are available. While using some blockchain networks, miners require a lot of processing power and high energy consumption. PoW and PoS are a few examples of consensus mechanisms, out of which especially proof of-, which is used by Bitcoin, consumes major energy resources and is a major source of carbon emission. Use of blockchain will raise the issues towards the environment also.

iv) **Concern of Interoperability**: Every blockchain platform works with its own protocols and standards, which will make it unique and strong for the fulfillment of various challenges of business economics. Data mapping from one blockchain to another is not possible due to the differences in their protocols. Interoperability across different blockchains is still difficult to achieve.

v) **Concern of Various Legislative Structures**: Very few of the jurisdictions have approved the cryptocurrency structure; but acceptance of technology is very high among them. Different jurisdictions have various blockchain and cryptocurrency regulatory environments. Adoption of blockchain technology for various domains will need to be mapped with the uncertainty of the laws of the structures' regulatory functions. Even the rules of decentralization are different among various regulatory laws which need to be positively incorporated.

To develop solutions to these problems, governments, businesses, academic institutions, and individuals must all collaborate. It entails creating laws and policies to protect user privacy, promoting universal access to digital resources and digital literacy, funding education and training for emerging job roles, fostering moral standards for technology development and encouraging responsible technology use to mitigate the negative social and psychological effects of technology.

Maximizing technology's good effects while actively addressing and minimizing any potential negative consequence is the ultimate goal as we work towards a more equitable and sustainable technological future.

4.2 BLOCKCHAIN BASICS AND ESSENTIALS

A blockchain is made up of many connected data blocks that are linked together in chronological order. The blockchain has many blocks. The very first block is called genesis block. It doesn't have the address of any other block. After the genesis block, each block contains the address of its previous block, so that the chain sequence can be maintained. Address is mapped in the form of a 256-bit hash code. Each block is a combination of block header and transactions. Block header consists of the block version, Merkle root, Timestamp, Nonce value, and the hash key of the previous block wherein a transaction contains the list of transactions and coin base.

Block version is a number which will help to track protocol upgrades. Merkle root is the topmost hash value of a transaction which is a combination of all the nodes at different levels. Merkle root is one of the most secure hash codes generated by the SHA-256 algorithm. Timestamp will tell the time of the block added to the blockchain. Nonce value is a value which always starts from zero and indicates increment to solve the mathematical puzzle for the verification of a transaction. Previous hash is the address of the previous block for the proper consecutive linking of blocks (Kang et al., 2018).

4.2.1 MANAGEMENT OF BLOCKS IN BLOCKCHAIN

Blockchain is the series of blocks consisting of multiple transactions in every block and all the blocks are interconnecting to each other for proper linking. Blockchain block management is dependent on a series of steps, each of which has multiple substeps. The first step is to choose the specific blockchain according to the requirement. Choosing the right blockchain for the right purpose will make the business more efficient and prompt. The basic understanding should be what kind of blockchain is required, i.e. private, public, or hybrid. The second step is the creation of a transaction and its broadcasting to the network after the validation process. The transaction management of each block is also one of the major aspects in validation of the blocks. Every block consists of one or multiple transactions which adds some of the major information in the block. The steps to be taken for managing blocks in blockchain in detail are explained below.

The process begins with the inception of a transaction that leads to creation of a new block. Transaction is the data to be added into blocks for the fulfillment of chains in blockchain. This transaction can be between two parties also where one party wants to forward some data or cryptocurrency to the other party. It can be a transfer of assets from one person to another or a digital money transfer in case of a crypto transaction. The transaction mainly leads to the data to be maintained in the blockchain. Once the transaction is done, the next phase is to look for the broadcast of the same into the network. Broadcasting the transaction requires the process of verification and validation (Mukherjee et al., 2021). The management of broadcasted transactions into the network is to maintain a peer-to-peer protocol, so that every node in the chain can have the information about the transaction and its management in blockchain. The verification part generally proceeds towards the checking of data viability, when it is transferred from one person to another. In the case of digital financial transactions, the main focus should be on whether the person/node who wants to transfer the digital asset has sufficient assets with him or not. For example, if a person wants to forward 100 Bitcoins to the other person using the Bitcoin blockchain, then the verification of the transaction should lead to checking whether the first person has 100 Bitcoins or not and the source of these 100 Bitcoins (Kamath Reshma, 2018).

Once the status of ownership of the funds is verified, then the next step is to look for a blockchain consensus mechanism. The consensus mechanisms are discussed later in the section 4.5. Different consensus mechanisms are used by various blockchain platforms. PoW is one of the first consensus mechanisms, used by miners to solve the mathematical puzzle for adding the block into the blockchain (Uddin et al., 2021). PoW is generally used by the Bitcoin platform of blockchain. Post that, the transaction is to be added into the blockchain with linking of the blocks from previously added blocks. The linking of blocks generally follows the concept of hash value of the previous block, so that the data from the past can also be fetched by linking them to the same block. Thus, adding a block in blockchain consists of creation of transaction/s, adding them in the block, encrypting the data, applying consensus mechanism for validation, and then adding the new block to the blockchain. Once the block has been added to the chain, the reflection of the block will be available to every peer of the blockchain.

The decentralized nature of blockchain, combined with cryptographic techniques and consensus mechanisms, makes it resilient against single points of failure and malicious attacks. This security and resilience are key factors in building trust within the network.

4.2.2 BLOCKCHAIN CORE COMPONENTS

Blocks play an important role in the world of blockchain. A block consists of a transaction or group of transactions. The core component of a transaction is data. The structure of blockchain depends on the organization and management of data. The primary data structure of a blockchain is the "chain of blocks,", which is immutable.

4.2.3 STRUCTURE OF BLOCKS IN BLOCKCHAIN

The structure of blocks in a blockchain depends on a few components namely hash value, Merkle Root, Merkle Tree, Cryptographic Algorithm, and Block Header. A blockchain consists of many interlinked blocks (AmponsahAnokye et al., 2021). The interlinking of blocks works on the concept of address mapping, i.e. hash value. The hashing process consists of converting data into an alphanumeric string by applying a cryptographic hash function. The SHA-256 algorithm (i.e. cryptographic hash function) applied to block data generates alphanumeric code of size 256 bits as hash value.

Basically, a block has two parts, block header and transactions. Block header is the main component for the linkage operation. The header contains metadata, timestamp of the block, and the hash value of the previous block. The hash value of the previous block will establish the linking between the previous block and current block, enabling data integrity. Thus, all other blocks in the blockchain will hold the address of the previous block except the very first block. The very first block in the blockchain, called genesis block, doesn't hold the address of any other block. The hash value of a block totally depends on the data inside it. Any small change in the data of a block will result in a totally different hash value. Therefore, after adding a block in blockchain, if one tries to make even a small change in the data of the block, it will lead to breaking of the link due to change in the hash value. Now, let's understand the generation of hash value in detail by using the concept of Merkle Tree and Merkle Root (Latifi Sobhan et al., 2019).

Figure 4.3 elaborates the connection between the blocks of a blockchain. The index of a block is the unique number representing a sequence in the blockchain making management and identification of the blocks in the blockchain easier.

Time Stamp: Time stamp is the date and time of adding the block to the chain. It can also be regarded as the time when the block was mined. In general, the mined time of the block and added time will be the same. Time stamp is very crucial for maintaining the sequential and progressive order and ensuring that blocks are added to the blockchain in a time-sequenced manner (Hueber O., 2019).

Hash Value: It is a unique value generated by applying SHA-256 algorithm. It is one kind of alphanumeric value depending on the data or transactions of the block. If the data inside the block are modified, the alphanumeric value will also get changed automatically. It is one of the components that is responsible for the security of data in the blockchain. More about hash value is discussed in the section 4.2.5 of the Chapter.

Previous Hash Value: It is the hash value of the block, previous to the current mined block and it is used for interlinking the previous block and current block.

Data: Data are a transaction or list of transactions mapped inside the block. Data are one of the major components in the block, for which the security concerns need to be addressed. For example, in the world of Bitcoin blockchain, the data would include transaction details, i.e. information of sender, information of recipient, and the total number of Bitcoins to be transferred. Figure 4.4 demonstrates the sample structure of the block of blockchain.

Now, let us understand the working of all the components in detail in the following sections 4.2.4 and 4.2.5.

4.2.4 MERKLE ROOT AND MERKLE TREE

The Merkle Root of a block is the final generated hash value for all the transactions in a block of a blockchain. The root node has many of the levels maintaining sub nodes. The Merkle Root node is calculated on the basis of the Merkle Tree. A tree structure data arrangement for which the combination of multiple hash values will generate a Merkle Root and that will be the final hash address given to the next node which is going to be added. Following is an example, illustrated in Figure 4.5, by which the concept of Merkle Root and Merkle Tree can be well understood.

The transactions of the block are the leaf nodes of the Merkle Tree. There are four transactions, so the Merkle Tree will have (4 − 1) i.e. 3 levels, and the individual transactions T1, T2, T3, and T4 will be at the 3rd level, represented as leaf nodes of

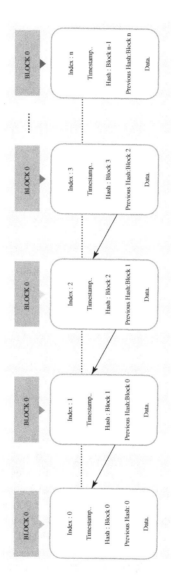

FIGURE 4.3 Genesis block: Architecture of blocks in blockchain. Source: Authors.

Block
{
   Index: 10412,
   Timestamp: 2023-08-05 10:40:56,
   Hash: 183b1a9a80faaf7581b692b884745d4f0c9af10b65789e7e8a9aeb1cc68c2312,
   Previous Hash: 4034de221b54cf2071d3f9212f01468c7a9fe365bc3f4c26e9415dfe2b381e2b,
   Data: { ... actual data ... }
}

FIGURE 4.4   Structure of data in a block.

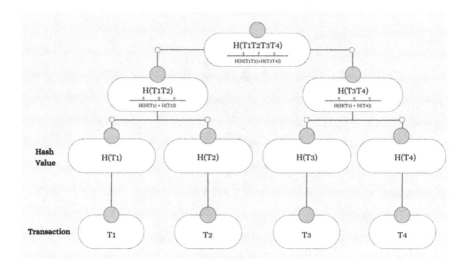

FIGURE 4.5   Merkle Root and Merkle Tree. Source: Authors.

the Merkle Tree. Let the hash values of these transactions generated be H(T1), H(T2), H(T3), and H(T4), respectively. Hash values of individual leaves will be represented at the 2nd level of the Merkle Tree, as parent of leaf nodes. Now a pair of two hash functions would be formed in the sequence and will be concatenated and then hash value again generated. That is, in this example H(T1) is concatenated with H(T2) and H(T3) is concatenated with H(T4) giving H(T1) + H(T2) and H(T3) + H(T4), respectively. Their hash values, i.e. H(H(T1) + H(T2) and H(H(T3) + H(T4)) are generated. H(H(T1) + H(T2) for brevity and convenience is expressed as H(T1T2); similarly H(H(T3) + H(T4)) is expressed as H(T3T4). H(T1T2) is the parent node of H(T1) and H(T2) whereas H(T3T4) is the parent node of H(T3) and H(T4) in the Merkle Tree at level one. Once again, this process is repeated, giving the root of the tree H(T1T2T3T4) at level 0, which is H(T1T2) + H(T3T4)). Root of this tree is Merkle Root (Chella et al., 2022).

In general, if a block has n transactions T1, T2, T3 … Tn, Merkle Tree will have $n - 1$ levels, with T1, T2, T3, … ,Tn as n leaf nodes at the nth level. Each of these leaf nodes will have a parent as H(Ti); i = 1, 2, 3,…,n at $(n-1)$ level. Then the pairs of nodes in sequence are concatenated, hash functions applied to concatenated string and at the next upper level, parent nodes are their hash function values. This process is continued, till one node is left, that is the root node, Merkle Root.

One of the well-known blockchains that uses SHA-256 is Bitcoin. SHA-256 algorithm is a hash function which generates 32-byte-long hash value. This, for our example is illustrated in Figure 4.5, each of H(T1), H(T2), H(T3), and H(T4) is 32 bytes long. This concatenation (H(T1) + H(T2) and H(T3) + H(T4) is of 64 bytes length and their parent nodes H(T1T2) and H(T3T4) is of 32 bytes. This way continuing Merkle Root is of 32 bytes.

4.2.5 SECURE HASH ALGORITHM (SHA-256)

SHA-256 is an abbreviation for a commonly used and extremely secure cryptographic hash function, short for Secure Hash Algorithm. It is useful for a variety of Computer Science applications such as password hashing and sensitive data hashing as well as digital signatures. Many blockchain applications, for example Bitcoin, Namecoin, Peercoin use SHA-256. SHA-256 is a part of the SHA-2 family of hash functions, designed by the National Security Agency (NSA) and published by the National Institute of Standards and Technology (NIST). SHA-256 takes an input message of arbitrary length and produces a fixed-size (256-bit) alphanumeric string as the output, known as the hash value (AkhigbeJoy, 2018). A few of the characteristics of SHA-256 algorithm are deterministic, Avalanche effect, irreversibility, and fixed-length alphanumeric code. Deterministic property means that the same hash value will be generated for a particular input. "Avalanche effect" means that a small change in the input data will produce an entirely different hash value. Irreversibility means no algorithm exists to produce original value from the hashed value. The length of the string produced by the SHA-256 is fixed to 256 bits, i.e. 32 bytes of alphanumeric code. It is a novel algorithm whose properties make it a crucial component in securing sensitive data and verifying the integrity of digital information.The design and security elements of the blockchain data structure, such as cryptographic hashing and block linking, make the system resistant to tampering and ensuring immutability (Rajnak et al., 2021).

4.3 TYPES OF BLOCKCHAIN

To meet the requirement of applications in different domains, different types of blockchain have been introduced. Different kinds of blockchain have different benefits for various situations depending on how they can be used. In general, blockchain can be classified into four categories, namely, Permissioned, Permissionless, Hybrid and Consortium, as illustrated in Figure 4.6.

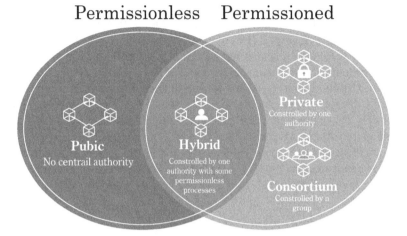

FIGURE 4.6   Types of blockchain. Source: Authors.

4.3.1 Permissioned Blockchain

Permissioned blockchain is a closed constrained network where the chain of blocks is in a restrictive environment of an individual or an organization. It follows the concept of DLT, i.e. distributed ledger technology, and manages the nodes in the peer-to-peer fashion. These blockchains are generally not very large because everyone on the internet is not permitted to join the network and participate in consensus mechanisms. Permissioned blockchains require the participants to obtain authorization or permission to access the network and engage in various activities. Private and Consortium type of blockchains come under the category of Permissioned blockchain. A private blockchain is controlled by a specific authority and the consortium is controlled by a group, responsible for granting access permission to a community (Mohanty Debajani 2019).

4.3.2 Private Blockchain

A private blockchain, also referred to as Permissioned blockchain, is generally a restricted blockchain designed inside the firewall of an organization, so that the restrictions can be managed more efficiently. Basically, private blockchains are designed for controlled environments where only known entities can collaboratively carry out the operations. Such a type of blockchains builds a higher level of trust among participants. Greater control over data protection, quicker processing of transactions, and customized governance models are all made possible by such types of restricted access. For the industries requiring security and confidentiality for their data, a private type of blockchain is more appropriate. The primary drawback of private blockchain is its small size, which may lower the security. Most common examples of private blockchain are Hyperledger Fabric and R3 Corda. Both of them are open source enterprise-grade permissioned distributed ledger technology (DLT) platforms (Amu D. and Santhi Baskaran 2022).

a) **Hyperledger Fabric**

Hyperledger Fabric is one of the open source projects of Hyperledger. As opposed to restricted domain-specific languages (DSL), Fabric is the first distributed ledger platform that supports smart contracts written in general-purpose programming languages like Go, Node.js and Java (Ana et al., 2018). This implies that further training of the individuals to learn a new language is not required as a majority of businesses are already working with such technologies and therefore possess the skill set required to create smart contracts. It also enables developers, teams and organizations to build together a few tools to simplify and secure the work. Healthcare is one of the industries, where Hyperledger Fabric can be used to develop a structure to maintain the digital records of an individual in sequence more securely. The documents can be fetched directly by having a unique id or a fingerprint, so that instant medication help can be easily provided to an individual.

b) **R3 Corda**

R3 Corda is an open source blockchain platform and is built on ISP principle, i.e. Interoperability–Security–Privacy principle. It is a suitable platform for financial institutions and enterprises. Corda can handle secured rapid execution, one of the most important needs of today's financial organizations. Corda is intended to provide privacy. It enables participants to share data transactions with only those individuals or organizations, who have authorization and access rights. This level of discretion is necessary for businesses working in sensitive financial and legal domains. Banking institutions intend to employ R3 Corda to keep global financial transactions agile and simple. According to the information provided by Andre Carneiro on bbchain .com, R3 Corda has been used to build applications for some of the world's largest institutions, including HSBC, ING, and J.P. Morgan (Khan et al.,2018).

4.3.3 CONSORTIUM BLOCKCHAIN

Consortium blockchain is generally a community blockchain. The community type stands for sharing of the same network by multiple trusted organizations for the completion of collective work. Each organization can enter into the dedicated network by having permission and work collectively. These types of blockchains are permissioned networks where no entity can enter without permission. Multiple organizations having rules and protocols to maintain and manage the data can use consortium blockchain for governance. MediLedger and IBM Food Trust are examples of consortium blockchain. MediLedger is basically designed for the healthcare industry globally. It is one of the major players in drug supply chain management, whereas IBM Food Trust, a product by IBM, is used for the improvisation of food supply chain management.

a) **MediLedger**

The healthcare business is one of the most important industries where people work collaboratively on the shared network to have the solutions of a particular problem. Collaborative work requires to establish protocols and business rules to control transactions and interactions among trading partners. MediLedger is one of the networks best suitable for enhancing the integrity and security of the pharmaceutical supply chain (Carolin Plewa et al., 2012). It was created to address challenges in the pharmaceutical industry such as traceability, counterfeit drugs, and regulatory compliance. MediLedger is also used to handle drug compliance and traceability.

b) **IBM Supply Chain Intelligence Suite – Food Trust**

IBM Food Trust program is a blockchain-based modular solution that benefits all network participants by making the food environment safer, smarter, and more sustainable. The IBM Food Trust is a collaborative supply chain network from beginning to end. This process includes seed distribution to farmers, crop management in farming, crop processing, crop cutting, and crop distribution to wholesalers and retailers. IBM Food Trust program oversees it all. This IBM blockchain-based solution connects parties through a permissioned and immutable blockchain.

4.3.4 PERMISSIONLESS BLOCKCHAIN

The terms "permissionless" and "public" bockchains are frequently used interchangeably. They share a key characteristic, i.e. open access to participants. It works on the concept of distributed ledger technology, where the nodes are available and accessible to everyone. Everyone with approved login on the blockchain is considered as a participant. The participants have rights to validate the transactions and are referred to as miners. Permissionless blockchains are typically unrestricted, simple to use, and regarded as the public variety of blockchains (Castro et al., 1999).

4.3.5 PUBLIC BLOCKCHAIN

This one allows you to write, view, and validate transactions without any limitations. These blockchains maintain a dispersed network of nodes. The security issues are taken care of by the use of a consensus process. Public blockchain are commonly used in a wide range of applications, including cryptocurrencies, smart contracts, and decentralized apps (Dapps) (Veronese et al., 2013). They are notable for their dependability and resistance to censorship, because no single entity controls the network. Despite being an open access blockchain, it is resistant to hacking as the security will be jeopardized only if a hacker gains control of more than 51% stake of blocks, which is difficult due to the size of blockchain. Bitcoin and Litecoin are two instances of the public blockchains, permitting participants to engage in activities without centralized control or permission.

a) **Bitcoin**: Bitcoin is the first and most well-known permissionless and public type of blockchain. It was introduced in a whitepaper published in 2008 by an individual or group using the pseudonym Satoshi Nakamoto. Bitcoin's principal function is to function as a decentralized digital currency, allowing peer-to-peer transactions without the use of intermediaries such as banks. It is one of the most popular cryptocurrencies in the world. The number of Bitcoin users has increased significantly over the years. The most significant advantage of Bitcoin is that it has been accepted and utilized not only by individuals, but also by institutions. Major corporations such as KFC and HGregoire, Canada, allow customers to purchase goods using Bitcoins (Armknecht et al., 2015).

b) **Litecoin:** A peer-to-peer internet currency called Litecoin offers quick, almost-free payments to anyone in the world. Litecoin is a fully decentralized, open source, worldwide payment network that lacks any central authorities. It is also permissionless, i.e. a public type of blockchain, often considered as a lighter version of Bitcoin. Litecoin is abbreviated as LTC and is a cryptocurrency that was created by Charlie Lee in October, 2011. It is considered as a "Silver to Bitcoin's Gold." It works on a similar technology, but the key distinction is that it confirms transactions more quickly as compared to Bitcoin. The transactions are confirmed four times faster than Bitcoin due to its Scrypt hashing algorithm. It is used by many of the individuals and communities for a variety of purposes. Many online merchants like Travala, eGifter, "Snel.com" and ShopinBit, accept the Litcoin (Abdelmaboud et al., 2022). Developers use Litcoins to create Dapps and tokens for the development of applications.

4.4   HYBRID BLOCKCHAIN

Hybrid blockchain is an accumulation of permissioned and permissionless type of blockchain, basically a combination of private and public blockchain. It offers a hybrid approach by allowing a few specific participants to access while keeping others in the restricted environment at the same time. Security measures or selective data management are the basic requirements of hybrid blockchain. The hybrid form of blockchain is helpful, if the requirement is to provide security or to have a specialized access to limited data management. Let's consider a scenario where the management wants to give access to all the employees of their internal application, but the data writing/modification rights will be provided to only a few managers. In this situation, a few portions of the network will be public, accessible to everyone and the rest will be private, restricted to only a few of the specific authorized persons. Thus, it blends the benefits of both types of blockchains. This technique aims to strike a balance between transparency and privacy, making it appropriate for applications requiring controlled data sharing and verification by various parties, while at the same time maintaining certain levels of anonymity. A few of the examples of hybrid types of blockchains are Dragonchain and Binance chain (Atlam et al., 2018).

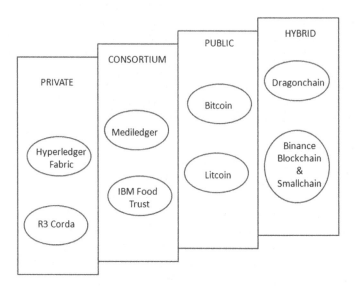

FIGURE 4.7   UseCases of blockchain types. Source: Authors.

a) **Dragonchain**: Dragonchain is an open source platform released in 2016. It was first developed by Disney as a private type of blockchain, but later converted into an open source platform to build and provide support to other applications also. It offers flexibility for the sharing of data and validations in a secured environment. By incorporating the advantages of blockchain technology with features that address realistic business needs, Dragonchain seeks to increase the usability and accessibility of blockchain technology for enterprises and developers.

b) **Binance Blockchain and Smart Chain**: Binance chain is a public chain used for cryptocurrency trade. Binance chain also provides a token known as Binance token. Binance smart chain is a sort of Permissioned blockchain, a parallel network to Binance chain, which, when combined with Binance blockchain, provides a hybrid environment (M. Samaniego, U. Jamsrandorj and R. Deters 2016). Binance provides two independents yet related blockchains. Binance chain serves as a decentralized marketplace for exchanging crypto assets, whereas Binance smart chain improves functionality by supporting smart contracts (Figure 4.7).

Table 4.1 illustrates the comparison of properties of different blockchain types in a nutshell.

4.5   CONSENSUS MECHANISM

The meaning of the word consensus is "agreement among a group of people." In blockchain technology consensus algorithms play a vital role. These are the

TABLE 4.1

Comparison of Blockchain Types

Properties		Types of Blockchain		
Types of Blockchain	Public	Private	Consortium	Hybrid
Aspect	Permissionless	Permissioned	Permissioned	Hybrid of Permissioned and Permissionless
Examples	Litecoin	Hyperledger Fabric	IBM Food Trust	Dragonchain
Governance	Decentralized	Centralized	Shared by a community	Combination of both
Security	Moderate	High	Balanced	Shared Security
Speed	Slow	Fast	Balanced	Balanced
Transparency	High	Low	Controlled	Balanced
Use Cases	Cryptocurrency	Internal Applications	Collaborative Applications	Custom
Readability	Open	Closed	Restricted to a community	Balanced

protocol-based algorithms designed specifically for fulfilling an agreement among network participants for the acceptability of a specific transaction to be added to the chain of blocks. Various blockchain platforms and technologies are paired with different consensus algorithms (Jafar Umar et al., 2021). Based on suitability, different blockchain platforms adopt a particular consensus algorithm according to their requirement. The following are a few major consensus algorithms used for adding the transaction into the specific blockchain. These are in general mathematical algorithms to solve a complex problem.

4.6   PROOF OF WORK (POW)

In the world of blockchain, the general abbreviation used for proof of work is PoW. It was first used by Satoshi Nakamoto in 2008. It is a consensus algorithm used by the Bitcoin platform. In this process, when an individual wants to add a block to the specific blockchain then it needs to get it verified by miners, who are responsible to validate the transaction by solving a complex mathematical problem. Many of the miners are allowed to participate in the process of validating a transaction but only the miner who will be able to solve the problem/mathematical puzzle first will get the authority to validate the transaction and right to add the block to the blockchain. Miners have to involve themselves in repeatedly hashing the block by a random value. The random value is called nonce value (Chen et al., 2019). The process will continue until the nonce value meets with the required criteria of satisfying the difficulty requirements. As a reward, the miner will get a few fractions of a Bitcoin, i.e. cryptocurrency, depending on the complexity of the problem solved. It is one of the initial consensus mechanisms used for validating a transaction. Bitcoin follows completely the PoW consensus mechanism/algorithm for adding a block into blockchain. The process of PoW requires high-end computational power which leads to high energy consumption and it is considered as one of the major drawbacks of the PoW consensus mechanism.

4.7   PROOF OF STAKE (POS)

The general abbreviation used for proof of stake is PoS and was developed by Sunny King and Scott Nadal in 2012. It is an alternative to PoW and aims to achieve similar objectives in a different way. In PoS, unlike PoW, not all the miners have to rush to solve the problem by using nonce value. In PoS the probability of choosing a miner is very high by applying a simple mechanism, i.e. stake or ownership of the cryptocurrency. Whoever has a high stake with them are first eligible for taking the part to identify the solution of the problem. This will filter all the miners at an early stage and would not keep all of them busy. The major benefit of PoS is that it is more energy-efficient than PoW and generally results in faster mining of blocks. Popular platforms like Ethereum, Polkadot, Avalanche, etc. use PoS (Yoshida H., Biryukov A. 2006). Each one of them is associated with a different variation of PoS, which leads them towards a faster rate of mining and less energy consumption. Ethereum earlier had association with PoW, but by 2022, Ethereum officially moved to PoS.

Because of the faster rate of mining, Ethereum is very much popular after Bitcoin in the current market and its token ETH is consistently growing into the market. But the problem with PoS is that the miner needs a large investment to maintain the stake.

4.8 DELEGATED PROOF OF STAKE (DPOS)

The general abbreviation used for Delegated Proof of Stake is DPoS and is developed by Daniel Larimer, founder of BitShares, Steemit, and EOS in 2014 [44]. It is one of the popular variants of PoS, and is designed to improve scalability and energy efficiency compared to traditional proof of work (PoW) systems like Bitcoin. It works on the voting system for tokens. In this system, token holders on the blockchain network have the ability to choose a predetermined number of delegates. These delegates can add new blocks to blockchain after validating the transactions. The number of delegates is typically not more than the total number of token holders in the network. In general, it is quite less compared to the number of token holders, allowing efficient and scalable consensus mechanism processes. It is a network algorithm, where only trusted validators have rights to validate the transactions. These rights are given by other network delegates of the specific blockchain. The major aim is to provide a more energy-efficient and scalable alternative to the traditional proof of work consensus mechanisms. DPoS aims to find a balance between decentralization, security, and efficiency by permitting all token holders to vote in the consensus process (Aithal et al., 2021).

4.9 PRACTICAL BYZANTINE FAULT TOLERANCE (PBFT)

The general abbreviation used for practical Byzantine fault tolerance is PBFT. Miguel Castro and Barbara Liskov came up with an idea through a study released in 1999 (Amiri et al., 2021). Only dependable validators can participate in a consensus algorithm for Permissioned blockchains. PBFT is a consensus algorithm for Permissioned blockchains that works with only trusted validators. It plays a major role in the world of internet of things (IoT) also. The major benefit of having the Byzantine fault tolerance is to manage a failure of the system. If a node is met with failure then the removal of node is suggested and optimizing the work in less time is the major usage of PBFT. It mainly offers the sequential order of all the nodes, where one node is the primary node and all other nodes are considered to be secondary nodes. PBFT mainly keeps the record so that no more than one-third of the total nodes have malicious activity. For example, if the total number of nodes is n, then not more than $n/3$ should have malicious activity. The practical Byzantine fault tolerance (PBFT) method is used by blockchain systems including Zilliqa, Hyperledger Fabric, and Tendermint.

4.10 PROOF OF AUTHORITY (POA)

The abbreviation for proof of authority is PoA and is a consensus mechanism with a limited number of trusted nodes. The major task of these trusted nodes is to validate a node for adding it into the blockchain. It is one of the best solutions used for private blockchain platforms. The major advantage of using PoA mechanism is that there is no need for high-end hardware and networking solutions to implement. This implementation leads to low power consumption and a faster rate of transaction confirmation. In PoA, there is a high degree of confidence and dependability among validators (Ncube et al., 2020). It is best suitable for consortium and private type of blockchains. However, PoA sacrifices the decentralization aspect found in PoW and many PoS systems, as the validation process is under the control of the authority nodes.

4.11   HYBRID APPROACH (PRACTICAL BYZANTINE FAULT TOLERANCE WITH PROOF OF STAKE)

This hybrid consensus mechanism combines the fault tolerance of PBFT with the energy efficiency of PoS. Validators are selected based on their stake in the network, and PBFT consensus is then used to finalize block agreement among these validators (Uddin Mueen, 2021). In our discussion of consensus mechanisms, we have noticed that choice of consensus mechanism will depend upon various factors like the decentralization level in the nodes of blockchain, security of blockchain, speed, scalability, storage, and energy efficiency. Table 4.2 gives a comparison of these consensus mechanisms in terms of different properties. One of the most important aspects to consider is the specific use case and domain of the blockchain network.

4.12   APPLICATIONS OF BLOCKCHAIN

Technology assists us in making our tasks easier and more efficient. Various applications of blockchain technology can be a part of our daily lives. Similarly, blockchain helps in various domains, i.e. RealEstate, Electronic Health Record System, Banking Transactions, Record Management System, and many more (Solat Siamak et al., 2020). These everyday applications can be made more safe and effective by utilizing blockchain technology. Based on the needs and specifications of implementing the system, any of the platforms including Bitcoin, Ethereum, Fatcom, R3 Corda, and Hyperledger can be chosen. The following is a discussion of a few of the applications of blockchain.

4.12.1   REAL ESTATE

The real estate market in India has the potential to get the economy back on track after Covid-19. The size of the Indian commercial real estate market USD 20.71 billion and the residential market size is expected to reach USD 182.41 billion in 2023. The process of buying or selling a property involves sale agreement or purchase contract, property disclosure documents, sale deed, i.e. ownership documents, mortgage documents, inspection/legal reports, and survey and the insurance certificate of the property buyer (Marar, Hazem and Marar, Rosana 2020). All these documents are

TABLE 4.2

Comparison of Consensus Mechanisms

Property	Consensus Mechanisms				
	PoW	PoS	DPoS	PBFT	PoA
Scalability	Low	Medium	High	Medium	High
Speed	Medium	Medium	High	High	High
Security	High	High	High	High	Medium
Storage (Decentralized)	High	Medium	Low	Low	Low
Energy Consumption	High	Low	Low	Low	Low
Participation	All Miners	Stakeholders	Token Holders	Nodes	Approved Entities
Examples	Bitcoin	Ethereum	Tron	Hyperledger Fabric	Quorum

considered to be important and the required documents for getting the sole ownership of the property and every document needs to be maintained for many years. Sometimes the property is transferred from one person to another person and due to this registration/ownership is transferred from one owner to another. The land titling and real estate industry stand to undergo a revolutionary transformation with the adoption of blockchain technology. The chain from the first person to the current owner needs to be maintained and blockchain technology can help it to be maintained securely and untampered. It is common to hear in some cases of the same property being sold to two or more people by a person, which is illegal. If the data are maintained in the blockchain, then the last legitimate operation is also assessed and verified. The verification of the transaction will increase security and control fraud activities in selling properties.

A permissioned or partially Permissioned blockchain is the need of the land titling and real estate industry, where the users should be trusted or known. Digitization of documents and choosing the right encryption algorithm is one of the basic requirements to get into the blockchain. Then the smart contracts need to be defined in such a way that they can elaborate the major terms and conditions for buying or selling the property. The smart contracts should be written perfectly so that the payment time period, historical ownership records, and property transfer terms' condition can be maintained.

The next step is tokenization of the property documents in the digital format. Consider the case where the property is owned by two or three individuals. In such a case, the respective tokens should be separately available to every owner of the property. An owner with partial ownership can surrender his/her tokens to allow another co-owner to sell the property. Also, tokenization of payment is required at the time of selling or buying the property. Further, the process of tokenization gives clarity in the taxation process while selling a property. This will lead to an advantage of managing non-fraudulent transactions. The challenges for the implementation of blockchain in this sector are collaboration among various departments, regulatory considerations, and overcoming technical maintenance.

4.12.2 Electronic Health Record System

An electronic health record (EHR) system utilizing blockchain technology can offer numerous benefits in terms of data security, privacy, interoperability, and patient empowerment. It is one of the growing sectors, where patients from other countries are visiting India. More than 800,000 foreigners visited India in the past three years for medical treatment, making it one of the top destinations for medical tourism. The Ministry of Home Affairs reported that between 2019 and 2021, over 825,000 medical visas were issued to foreigners, and majority of them were from African countries (Alexander et al., 2022).

India is a huge market for the pharmaceutical sector. Many patients are traveling to India for medical treatment and recuperation. The Indian health care sector has a lot of potential in the future. The implementation can be done from two aspects.

i) The first aspect is that every medicine is a combination of a few drugs. That combination was specified at the time of the drug's invention. It cannot be changed or amended after that. In this situation, blockchain technology can be used to store the combination of the drug, making the combination once stored non-mutable. Then, on a visit to the doctor, the doctor must prescribe only the drug combinations to the patient, and the patient may purchase any brand medicine associated with the drug or combination of the drug (Lashkari Bahareh and Musilek Petr, 2021).

ii) Another aspect is to maintain individual health care data in the series of chains of blocks which can be unlocked by using fingerprint scanners. Every time one visits a doctor, the number of files (i.e. in case of visiting the same doctor with whom one has consulted earlier) or a prescription (i.e. in the case of a new doctor) will be added. The digitization of every health document is the need of the hour, where the documents can be digitized and unlocked either by using the Aadhar card number or fingerprint. It will save the process of maintaining unnecessary hard copies of documents. The doctor can also identify the allergies of a specific drug and other diseases by assessing the patient data. Consider a scenario where a person has met with an accident and is not in a condition to speak. Now at this stage, the doctor can use his/her digitized profile or data unlocked by a fingerprint scanner and start the basic treatment which can save a valuable human life.

4.12.3 BANKING TRANSACTIONS MANAGEMENT

Blockchain technology is expected to change conventional financial service business models. The major players in conventional financial services are private and public banking institutions (Nguyen et al., 2019). Transaction management in banks requires confidentiality, privacy, and integrity. A transaction between two parties should follow all the rules of privacy management. In the world of blockchain, R3 Corda allows the management of transactions with privacy. It is an open source technology implemented using Java and Kotlin. JVM plays a major role in the implementation of any business by using R3 Corda. It is a distributed ledger technology developed in the year 2014 by David E. Rutter. The major benefits of using R3 Corda is that it is considered as an Open Permissioned blockchain, where the data are shared to only legitimate parties rather than sharing them with the whole network.

As the digital assets are increasing day by day, banks have accepted the solutions to maintain the transactions in the permanent and non-tampered chain of blocks. R3 Corda is an enterprise ready open source platform which can be used and even modified according to one's requirements, increasing the acceptability and adaptability of the corda technology among various industries. It is a shared ledger technology and is used by many of the major finance giants to streamline their operations and manage the cost. According to Andre Carneiro, R3 Corda is adapted by a few of the world's leading banks to reduce the cost and maintain private transactions. The major banks HSBC, J.P. Morgan, etc. have built their own customized application on the basic concepts of R3 Corda. They use a vault database to maintain the local data transaction between two parties. The IOU model records how much money is owed by a person to another person. Corda also supports legal compliances of auditing and functioning towards identity management, which will support the business.

4.12.4 CROSS-BORDER PAYMENT SERVICES

People migrating to other countries for better opportunities are very common nowadays. As a result, the need for global money transactions has increased. According to the *Economic Times*, with the Indian diaspora growing around the world, India was the top beneficiary of remittances in 2022, receiving $89,127 million. The United States of America and the United Arab Emirates cumulatively contributed to over 40% of the inward remittances at 23.4% and 18%, respectively (Bertoni et al., 2013).

Sending money from one country to another is a common practice, especially to one's native country. Instant money transfer does not take place when people from abroad send money to their home country. The major problem here is the time taken for the money to reach the recipient. The speed and efficiency of this process are affected by a number of factors, including the mode of transfer, the countries involved, and the financial institutions or services used. Money transfer is not a hazardous process, but the basic problem is that the sender bank does not have an account in the receiver country and due to this reason, they have to use third-party services for mapping the transaction. Also, there is a processing cost attached to the transfer of money. The process is clumsy, slow, and non-transparent. Ripple is the solution to such problems, it is a real-time gross settlement system by which it is possible to exchange the currency inter-country at a very faster rate and low cost. It was created by Ripple Labs Inc. in the year 2012. The main benefit of having Ripple is that it offers fast and secured cross-border payments by using Ripple X current messaging technology. More than 100 banks are using the X current messaging services offered by Ripple (Morris, David Z., 2016). XRP is a cryptocurrency introduced by Ripple. However, banks have refused to accept it since they do not prefer to maintain their money as virtual currency by using blockchain.

4.12.5 DEMURRAGE CURRENCIES IN BLOCKCHAIN

Demurrage currencies are the ones that focus on the flow of currency in the market. It works on the concept of encouraging spending money and restricts the model of holding the money. Holding the money incurs a cost, which will be borne by the individual who has held it. The basic aim is to promote the flow of money in the economy by incentivizing people to put their money to use instead of holding onto it. In the case of non-usage of money, the money will depreciate by any fixed value over a period of time. Blockchain can improvise the concept of Demurrage currencies, by using smart contracts and controlling the expiry of the currency. Let us take an example, where the government has given some amount for a specific work to a private agency with a commitment to complete the work in a specific time period. For the same, the government has also provided upfront money to the private agency.

After the specific time period, if the work has not been completed then the remaining money starts depreciating. The impact of that will be on the completion of work in time. Few of the currencies like Earth Dollars, Freicoin, and Mutual credit are working on the concept of Demurrage (Popper & Nathan, 2016).

4.12.6 BLOCKCHAIN-ENABLED IoT APPLICATIONS

The internet of things (IoT) has gained popularity as a paradigm for computing technology. IoT applications are the requirement of today's world, where multiple devices are connected on a network for the ease of usage. With the feasibility of connecting multiple devices with each other through the internet, the IoT market is expanding at an unexceptional rate. Blockchain technology has a potential to map the world of IoT in a secure mechanism. The data on the internet has a few primary challenges, which are security, efficient mapping of data, interoperability, and scalability. Resource sharing and scalability are becoming important factors because of the volume and velocity of data on the internet. The world of IoT has millions of devices to connect in the same network, resulting in delayed data processing.

All of these challenges can be addressed with blockchain technology by managing networked devices decentralized, and high scalability can be accomplished due to the blockchain's rapid consensus mechanism. Blockchain helps IoT devices to communicate directly with each other without following the concept of any centralized authority. This decentralized approach enhances trust and reliability, as there is no single point of failure or vulnerability in the system. The other aspect is security, when all the IoT devices are in the open network. Blockchain has the capacity to enable secure identity management for such devices. With the help of cryptographic algorithms only authorized devices can interact with the network and perform specific actions. Blockchain integration with IoT devices has begun to spread among many businesses. IBM and Samsung have collaborated to create the autonomous decentralized peer-to-peer telemetry (ADEPT), a platform that exploits components of Bitcoins for building a decentralized IoT network of devices.

4.12.7 BLOCKCHAIN-BASED SECURED ELECTRONIC VOTING SYSTEM

Elections are an essential component of a democratic society because they allow the general public to express their opinions by voting. Since 1970, electronic voting, or e-voting, has been accomplished through a variety of methods. Many systems created specifically for voting have considerable advantages over paper-based systems, including increased efficacy and lower errors. Online voting is another medium, where voting by email is accomplished in some circumstances, such as absentee voting, which allows one to vote before Election Day via email or drop box. Online voting is becoming more popular in modern society. It has the ability to reduce organizational costs while increasing voter turnout. There are numerous alternatives to email voting accessible under the umbrella of online voting. Various technologies have been used for the development of such systems resulting in the online secure voting system (Jamader, A. R., Das, P., & Acharya, B., 2022). The primary challenges that need to be mapped here are transparency, privacy, and tampering of data. More issues with an online voting system is to maintain the record of the votes, managing of votes received from different locations, and identification of authentic votes. Authentic vote means that the existence of the voter needs to be verified, so that fake votes can be easily identified. This system needs collaboration with other government databases also and must keep the identity of the voter disclosed from the party. Blockchain technology has the potential to address all the issues and make online electronic voting systems more strong. A blockchain-enabled system can be developed by using decentralized and immutable ledger to record and verify votes. After successful voting, miners shall validate and add the transactions into the public ledger. This will lead to faster, secured, and non-tampered processing of all the votes. On the basis of data, major analysis of non-voting areas can be also done. Using the online e-voting approaches the voting ratio is definitely going to increase worldwide (Jamader et al., 2023).

4.12.8 FUTURE SCOPE OF BLOCKCHAIN

The potential applications and implementation sectors for blockchain technology are numerous and diverse. Blockchain is anticipated to have a substantial impact on a number of crucial industries, including digital content management and record keeping, Intellectual Property Rights, supply chain management for various industries, academia, and many more. The future is IT-enabled, with numerous services and apps. Today, people are using many of the applications for a variety of purposes. Sharing our personal resources on the internet is common nowadays. People share their pictures, locations, and their hobbies frequently on the internet via various platforms. Many platforms will integrate in the future, providing benefits in a variety of ways. Collaborative applications are always beneficial for IoT-enabled applications. Security is a crucial issue for the future, which blockchain technology can address in all of these applications. By utilizing the notion of smart contracts in blockchain technology, financial and legal applications can establish their agreements and perform the execution without the use of third-party intermediaries such as banks and lawyers. Blockchain consensus mechanisms are perfectly designed, which can lower the cost and transaction time (Arayanan et al., 2016). The optimal architecture of the blockchain consensus mechanism can reduce transaction costs and times. The concept of cloud-computing platforms is used by the majority of industries. Blockchain technology is one of the perfect suitable solutions to be integrated with cloud computing in several ways to enhance security, transparency, and efficiency. Despite the great potential of blockchain technology, widespread adoption is still a work in progress. Energy consumption and legal frameworks with existing systems are all the issues that must be addressed before blockchain can reach its full potential in these diverse fields (Jamader, A. R., Chowdhary, S., & Shankar Jha, S. 2023).

4.13 CONCLUSION

Blockchain technology has emerged as a game-changing breakthrough with the ability to revamp several processes and industries. Distributed ledger technology and non-mutability are a few of the features that have altered people's perceptions about blockchain technology and moved them towards its acceptance. Though, the road to mainstream adoption is not without challenges. Legal compliance, scalability, energy efficiency, and seamless integration must all be balanced for blockchain to realize its full potential. Collaboration among developers, governments, various businesses, and services will be critical in defining how blockchain will impact the future of handling data efficiently and effectively with increased security. This chapter focused on numerous blockchain applications that have the potential of transforming society. As digital data and applications are growing exponentially, the management of these data and applications can be done efficiently through blockchain technology, which will be beneficial to all types of organizations.

Blockchain has the potential to accelerate the business applications in reshaping the future ecosystem of digital India. Thus, the adoption of blockchain technology is in its initial phases, and the benefits of adopting it cannot be ignored. It has a great future ahead. Soon, enterprises will be using it for effective and secure business execution as well as to mark their position in a competitive marketplace.

REFERENCES

Akhtar, F., Li, J., Heyat, M.B.B., Quadri, S., Ahmed, S., Yun, X., and Haq, A. (2019), "Potential of Blockchain Technology in Digital Currency: A Review", 16th *International Computer Conference on Wavelet Active Media Technology and Information Processing (ICCWAMTIP)*, pp. 85–91, doihttps://doi.org/10.1109/ICCWAMTIP47768.2019.9067546.

Amponsah, A., Adekoya, A., and Benjamin, W. (2021), "Blockchain in Insurance: Exploratory Analysis of Prospects and Threats", *International Journal of Advanced Computer Science and Applications*. Vol 12, Issue 1, doihttps://doi.org/10.14569/IJACSA.2021.0120153.

Atlam, H., Alenezi, A., and AlassafiMadini, W.G. (2018), "Blockchain with Internet of Things: Benefits, Challenges and Future Directions", *International Journal of Intelligent Systems and Applications*, Online ISSN: 2074–9058, https://doi.org/10.10.5815/ijisa.2018.06.05.

Akhigbe, J. (2018), "The Future of Crypto-Currency in the Absence of Regulation, Social and Legal Impact", *International Journal of Social Sciences*, Vol 4, Issue 1, pp. 555–570, https://doi.org/10.20319/pijss.2018.41.555570.

Amu, D. and Santhi, B. (2022), "A Survey of Applications Using Blockchain Technology", *International Conference on Computer Communication and Informatics (ICCCI)*, Coimbatore, India, ISSN: 2329–7190, pp. 1–6, 2022, https://doi.org/10.1109/ICCCI54379.2022.9740958.

Abdelmaboud, A., Ahmed, A.I.A., Abaker, M., Eisa, T.A.E., Albasheer, H., Ghorashi, S.A. and Karim, F.K. (2022), "Blockchain for IoT Applications: Taxonomy, Platforms, Recent Advances", *Challenges and Future Research Directions, Electronics*. Vol 11, Issue 4. IEEE Smart Data, pp. 433–436, 2016, https://doi.org/10.3390/electronics11040630.

Aithal, P.K., Saavedra, P.S., Aithal, S. and Ghoash, S. (2021), "Blockchain Technology and its Types-A Short Review", *International Journal of Applied Science and Engineering*. Online ISSN: 2322–0465, Vol 9, pp. 189–200.

Alexander, C., Heck, D. and Kaeck, A. (2022), "The Role of Binance in Bitcoin Volatility Transmission", *Applied Mathematical Finance*. Vol 29, pp. 1–32. https://doi.org/10.1080/1350486X.2022.2125885.

Arayanan, A., Bonneau, J., Felten, E., Miller, A., and Goldfeder, S. (2016), *Bitcoin and cryptocurrency technologies: a comprehensive introduction*. Princeton, New Jersey: Princeton University Press. ISBN 978-0-691-17169-2. Retrieved on September 7, 2023

Ana, R., Cristian, M., Jaime, C., Enrique, S. and Manuel D. (2018), "On Blockchain and Its Integration with IoT Challenges and Opportunities", *Future Generation Computer Systems*, ISSN 0167-739X, Vol 88, pp. 173–190, https://doi.org/10.1016/j.future.2018.05 .046.

Armknecht, F., Karame, G.O., Mandal, A., Youssef, F. and Zenner, E. (2015), "Ripple: Overview and Outlook", *Trust and Trustworthy Computing, Lecture Notes in Computer Science*. Vol 9229, Springer, Cham. https://doi.org/org/10.1007/978 -3-319 -22846-4_10.

Bansod, S. and Ragha, L. (2022), "Challenges in Making Blockchain Privacy Compliant for the Digital World: Some Measures", *Sādhanā*, ISSN: 0973–7677, Vol 47, Issue 3, p. 168, https://doi.org/10.1007/s12046-022-01931–1.

Bertoni, G., Daemen, J., Peeters, M., and Van Assche, G. (2013), Keccak, Advances in Cryptology – EUROCRYPT 2013. EUROCRYPT 2013. *Lecture Notes in Computer Science*, vol 7881. Springer, Berlin, Heidelberg. https://doi.org/10.1007/978-3-642 -38348-9_19.

Chen, Y.-C., Chou, Y.-P. and Chou, Y.-C. (2019), "An Image Authentication Scheme Using Merkle Tree Mechanisms", *Future Internet*, Online ISSN: 1999–5903, Vol 11, p. 149. https://doi.org/10.3390/fi11070149.

Castro, M. and Liskov, B. (1999), "Practical Byzantine Fault Tolerance", *Proceedings of the Third Symposium on Operating Systems Design and Implementation*, New Orleans, USA, ISBN: 978-1-880446-39-3, pp. 173–186.

Carolin, P., Indrit, T., Anthony, F. and Giselle, R. (2012), "Technology Adoption and Performance Impact in Innovation Domains", *Industrial Management and Data Systems*, ISSN 0263–5577, Vol 112 Issue 5, pp. 748–765. https://doi.org/10.1108 /02635571211232316.

Chelladurai, U. and Pandian, S. (2022), "A Novel Blockchain Based Electronic Health Record Automation System for Healthcare", *Journal of Ambient Intelligence and Humanized Computing*, Online ISSN: 1868–5145, Issue 13. https://doi.org/10.1007/ s12652-02103163–3.

Veronese, G.S., Correia, M., Bessani, A.N., Lung, L.C. and Verissimo, P. (2013), "Efficient Byzantine Fault-Tolerance," in *IEEE Transactions on Computers*, Vol 62, no. 1, pp. 16–30, https://doi.org/10.1109/TC.2011.221.

Hueber, O. (2019), "Sidechain and Volatility of Cryptocurrencies Based on the Blockchain Technology", *International Journal of Community Currency Research*, ISSN 1325– 9547, Vol 23, Issue 2, pp. 35–44, https://doi.org/10.15133/j.ijccr.20 19.012.

Jafar, U., Aziz, M.J. and Shukur, Z. (2021), "Blockchain for Electronic Voting System-Review and Open Research Challenges", *Sensors (Basel)*, Vol. 21, Issue 17, p. 5874. https://doi.org/10.3390/s21175874.

Jamader, A.R., Chowdhary, S. and Shankar Jha, S. (2023). A Road Map for Two Decades of Sustainable Tourism Development Framework. In *Resilient and Sustainable Destinations After Disaster: Challenges and Strategies* (pp. 9–18). Emerald Publishing Limited.

Jamader, A.R., Chowdhary, S., Jha, S.S. and Roy, B. (2023). "Application of Economic Models to Green Circumstance for Management of Littoral Area: A Sustainable Tourism Arrangement", *SMART Journal of Business Management Studies*. Vol 19, Issue 1, pp. 70–84.

Jamader, A.R., Das, P. and Acharya, B. (2022). An Analysis of Consumers Acceptance towards Usage of Digital Payment System, Fintech and CBDC. *Fintech and CBDC (January 1, 2022)*.

Krstić, M. and Krstić, L. (2020), "Hyperledger Frameworks with a Special Focus on Hyperledger Fabric", *Vojnotehnickiglasnik*. Vol 68, Issue 3, pp. 639–663, https://doi .org/10.5937/vojtehg6826206.

Kang, J., Xiong, Z., Niyato, D., Ye, D., Kim, D.I. and Zhao, J. (2018), "Towards Secure Blockchain-enabled Internet of Vehicles: Optimizing Consensus Management Using Reputation and Contract Theory", *IEEE Transactions on Vehicular Technology*, Online ISSN: 1939–9359, Vol 99, pp. 1–1, https://doi.org/10.1109/TVT.2019.2894944.

Kamath, R. (2018), "Food Traceability on Blockchain: Walmart's Pork and Mango Pilots with IBM", *The Journal of the British Blockchain Association*, Online ISSN: 2516–3957, Vol 1, Issue 1, pp. 1–12. https://doi.org/10.31585/jbba-1-1-(10).

Khan, K., Arshad, J. and Khan, M. (2018), "Secure Digital Voting System Based on Blockchain Technology", *International Journal of Electronic Government Research (IJEGR)*, ISSN: 1548–3886, Vol 14, Issue 1, https://doi.org/10.4018/IJEGR.2018010103.

Lashkari, B. and Musilek, P. (2021), "A Comprehensive Review of Blockchain Consensus Mechanisms", *IEEE Access*, Online ISSN: 2169–3536, Vol 4, pp. 1–1. https://doi.org/10 .1109/ACCESS.2021.3065880.

Latifi, S., Zhang, Y. and Cheng, L.-C. (2019), "Blockchain-Based Real Estate Market: One Method for Applying Blockchain Technology in Commercial Real Estate Market", *2019 IEEE International Conference on Blockchain (Blockchain)*, Atlanta, GA, USA, 2019, pp. 528–535, https://doi.org/10.1109/Blockchain.2019.00002.

Miglietti, C., Kubosova, Z. and Škuláňová, N. (2019), "Bitcoin, Litecoin and the Euro: An Annualized Volatility Analysis", *Studies in Economics and Finance*, https://doi.org/10 .1108/SEF-02-2019–0050.

Mukherjee, P. and Pradhan, C. (2021), "Blockchain 1.0 to Blockchain 4.0. The Evolutionary Transformation of Blockchain Technology", In *Blockchain Technology: Applications and Challenges*, Springer International Publishing, ISBN: 9873030693954, https://doi .org/10.1007/978-3-030- 69395-4_3.

Mohanty, D. (2019), "R3 Corda for Architects and Developers: With Case Studies in Finance, Insurance, Healthcare, Travel, Telecom and Agriculture", ISBN: 978-1-4842- 4531–6, https://doi.org/10.1007/978-1-4842-4529–3.

Samaniego, M., Jamsrandorj, U. and Deters, R. (2016), "Blockchain as a Service for IoT", *IEEE International Conference on Internet of Things (iThings) and IEEE Green Computing and Communications (GreenCom) and IEEE Cyber, Physical and Social Computing (CPSCom) and IEEE Smart Data (SmartData)*, Chengdu, China, 2016, pp. 433–436, https://doi.org/ 10.1109/iThings-GreenCom-CPSCom-SmartData.2016.102.

Marar, H. and Marar, R. (2020), "Hybrid Blockchain", *Jordanian Journal of Computers and Information Technology*. Vol 6, Issue 4, p. 1. https://doi.org/10.5455/jjcit.71–1589089941.

Morris, D.Z. (2016, May 15). "Leaderless, Blockchain-Based Venture Capital Fund Raises $100 Million and Counting", *Fortune*. Archived from the original on 21 May 2016. Retrieved on September 7, 2023.

Nguyen, C., Dinh, T.H., Nguyen, D., Niyato, D., Nguyen, H. and Dutkiewicz, E. (2019), "Proof-of-Stake Consensus Mechanisms for Future Blockchain Networks: Fundamentals, Applications and Opportunities", *IEEE Access*. ISSN: 2169–3536, pp. 1–1. https://doi .org/10.1109/ACCESS.2 019.2925010.

Samsir, N.A. and Jain, A.A. (2022), "Comparison of Blockchain platforms for the Implementation of Healthcare Applications", *High Technology Letters*, ISSN: 1006– 6748, Vol 28, Issue 8.

Ncube, T., Dlodlo, N. and Terzoli, A. (2020), "Private Blockchain Networks: A Solution for Data Privacy", *2020 2nd International Multidisciplinary Information Technology and Engineering Conference (IMITEC)*, Kimberley, South Africa, pp. 1–8, https://doi.org /10.1109/IMITEC50163.2020.9334132.

Samsir, N.A. and Jain, A.A. (2021), "Overview of Blockchain Technology and Its Application in Healthcare Sector", *Sixth International Conference on Information and Communication Technology for Competitive Strategies (ICTCS 2021) Lecture Notes in Networks and Systems*, vol 400. Springer, Singapore.

Nakamoto, S. (2008), Bitcoin: A Peer-to-Peer Electronic Cash System. https://bitcoin.org/bitcoin.pdf.

Samsir, N.A. and Jain, A.A. (2022), "Understanding and Comparative Analysis of Consensus Algorithms", *7th International Conference on ICT for Sustainable Development (ICT4SD 2022), Lecture Notes in Networks and Systems*, Springer, Singapore. ISSN: 2367–3370, Series: https://www.springer.com/series/15179.

Panda, S., Jena, A., Swain, S. and Satapathy, S. (2021), "Blockchain Technology: Applications and Challenges", In *Intelligent Systems Reference Library*, Vol 203. Springer ISBN:978-3-030-69395-4, https://doi.org/10.1007/978 -3-030-69395-4.

Popper, N. (2016, May 21). "A Venture Fund with Plenty of Virtual Capital, but No Capitalist", *The New York Times*. Archived from the original on 22 May 2016. Retrieved on September 7, 2023.

Rajnak, V. and Puschmann, T. (2021), "The Impact of Blockchain on Business Models in Banking", *Information Systems and e-Business Management*. Online ISSN: 1617–9854, Issue 19. https://doi.org/10.1007/s10257-020- 00468-2.

Haber, S. and Stornetta, W.S. (1991), "How to Time-Stamp a Digital Document", *Journal of Cryptography, International Association for Cryptographic Research Lecture*, Vol 537, pp. 99–111. Springer, Berlin, Heidelberg, https://doi.org/10.1007/3-540-38424-3 _32.

Solat, S., Calvez, P. and Naït-Abdesselam, F. (2020), "Permissioned vs. Permissionless Blockchain: How and Why There Is Only One Right Choice", *Journal of Software*, ISSN: 1796-217X, Vol 16, Issue 3, pp. 95–106. https://doi.org/10.17706/jsw.16.3.95–106.

Uddin, M. (2021), "Blockchain Medledger: A Hyperledger Fabric Enabled Drug Traceability System for Counterfeit Drugs in Pharmaceutical Industry", *International Journal of Pharmaceutics*. ISSN 0378–5173, Vol 597, p. 120235. https://doi.org/10.1016/j.ijpharm .2021.120235.

Uddin, M., Salah, K., Jayaraman, R.P. and Ellahham, S. (2021), "Blockchain for Drug Traceability: Architectures and Open Challenges", *Health Informatics Journal*. Print ISSN: 1460–4582, Vol. 27. https://doi.org/10.1177/14604582211011228.

Yoshida, H. and Biryukov, A. (2006), "Analysis of a SHA-256 Variant", *12th international workshop on Selected Areas in Cryptography. SAC 2005. Lecture Notes in Computer Science*, Vol 3897. Springer, Berlin, ISBN: 978-3-540-33109-4, Heidelberg. https://doi .org/10.1007/11693383_17.

5 Empowering Energy Transition through Blockchain Technology

Oum Kumari R.

5.1 INTRODUCTION

According to the UNEP report 2022, the top seven countries—China, USA, India, EU, Russia, Indonesia, and Brazil—account for 50% of global greenhouse gas emissions (UNEP 2022). Almost all countries have announced net-zero targets to be achieved 2050–2070. Energy sector alone contributes three-fourth of the total greenhouse gases (IEA 2021) and effective energy transition towards renewable energy sources like solar, wind, etc., can only help us reduce the emission and achieve net-zero targets worldwide. Traditional centralized grid systems may not help the economies in achieving net-zero targets as they are suitable for economies relying on fossil fuels for their power demand. Net-zero targets can be achieved with the help of a flexible grid, which can easily adapt to the dynamic nature of renewable energy supply and demand. Nations are striving hard to attain the status of "Developed Nation," which has led to an increase in the production and consumption of energy in the last few decades (Oum Kumari, 2019). The ever-increasing demand for energy has forced the sector to increase the installed capacity to strike a balance between demand and supply. The per capita of energy consumption of the world from primary sources has increased from 27,972 TWh in 1950 to 160,764 TWh in 2022. Fossil fuels have continued to dominate the energy sector for the last two centuries and the dependence on fossil fuels for energy is shown in Figure 5.1.

Heavy reliance on coal, oil, gas, and traditional biomass has led to unsustainable production and consumption of natural resources, which has caused climate change and global warming (Khan et al., 2022). Sustainability includes three dimensions of development i.e. social, economic, and environment (Liu., 2022), and adoption of the precautionary principle for social, economic, and environment problems would help nations achieve higher growth without harming the planet (Taghizedh F., et al., 2022). Nations have been striving hard to strengthen their economies through adequate energy supply, but the price paid in terms of environmental degradation is very high. Affordable energy and a low-carbon future have now become the primary objectives of nations and blockchain technology is considered a powerful tool to accelerate and empower the energy transition process through a decentralized

DOI: 10.1201/9781003453109-5

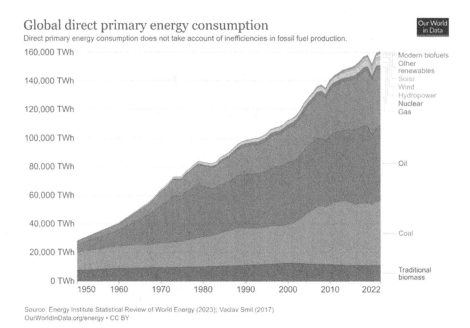

FIGURE 5.1   Primary energy consumption. Source: Author.

energy ecosystem. The present chapter discusses the integration of energy and economic development; the use of blockchain technology in the energy sector; and the challenges of BCT in energy transition, along with the discussion and conclusion.

5.2   ENERGY AND ECONOMIC DEVELOPMENT

Energy plays a vital role in human lives and the growth of a nation, and energy shortage hinders both human and economic development (HuoJie et al., 2013). Energy consumption and economic development have a strong positive correlation with one another, as shown in Figure 5.2. A high rate of growth and development calls for higher per capita consumption of energy, and higher consumption of energy calls for the development of any nation.

According to Energy Outlook, 2019, the demand for energy at a global level is expected to increase by 30% to 35% in 2040 due to an increase in the standard of living of people and reduction in poverty (Wen et al., 2021), mostly in developed and developing countries like India and China, which are densely populated (Chen et al., 2021). This again calls for serious actions against climate change as more than 60% of energy demand is met through fossil fuels, resulting in the emission of Greenhouse Gases. The development of a nation is not possible without achieving energy security, as the process of industrialization, standard of living, infrastructural development, etc., depends on an uninterrupted, affordable energy

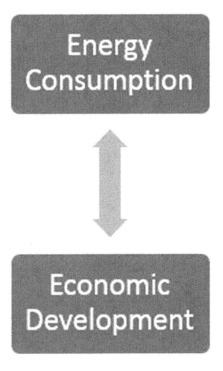

FIGURE 5.2 Energy consumption and economic development. Source: Author.

supply across the nations. Energy security, i.e. a nation's ability to provide sustainable energy resources without affecting its growth and development, is considered to be an effective parameter determining the nation's growth (Chi-Chuan, 2022). Most of the nations across the globe are tied to a centralized grid system to deliver electricity from power plants to ensure a reliable and efficient supply of power. However, the centralized grid system has resulted in various other challenges such as huge energy losses due to resistance in transmission lines, transformer losses, corona discharge, theft, and distribution losses. Power generated from fossil fuels creates high negative externalities; therefore, an energy transition towards renewables along with a decentralized energy supply system can help mitigate problems of energy deficit, environmental degradation, and accelerate the economic growth of the nation.

5.3 METHODOLOGY

The present study is based on existing literature on blockchain technology and its application in the energy sector. The study has also considered successful cases from other developed countries such as the USA, UK, and China. The data used in the study is collected from articles, internet sources, and published articles.

5.4   BLOCKCHAIN TECHNOLOGY AND ENERGY TRANSITION

The energy sector is undergoing a serious transformation through blockchain technology (BCT). The communication between energy resources, solar technologies, net metering systems, and smart meters can be facilitated by BCT. Optimizing the use of renewables can benefit the economy by helping overcome the energy crisis and reduce the emission of greenhouse gases effectively. Technological advancements during the last two decades have improved the performance of the energy sector across the globe. However, the centralized model of energy has not been successful in overcoming the energy issues in developing nations. BCT can transform the centralized energy sector into a more sustainable and decentralized model, helping the nation to tap the potential of green energy sources and solve energy and environmental issues simultaneously. All the developing and developed economies across the globe have announced net-zero targets by 2050 and 2070. BCT technology can help nations to explore renewables and optimize their use, thereby making our planet a better place to live without compromising growth. Security and transparency, being the significant features of BCT allows efficient distribution of energy, thereby reducing the cost of distribution.

Blockchain technology, a distributed ledger, is the technology behind cryptocurrencies and can also can play a key role in the process of energy transition. BCT has the potential to achieve the Paris climate goal of 1.5C towards the end of the 21st century by enabling us to adopt renewables and reduce our dependence on coal and oil, thereby cooling our planet. BCT can help nations to accelerate the energy transition process towards renewables through digitalization and decentralization (IFPEN). Currently, economies meet their increasing energy demand by using fossil fuels especially coal and oil. The continuous increase in per capita consumption of power in countries has led to an increase in greenhouse gas emissions and mounting debts, as most of these energy sources are imported from other countries. (IEA). The existing infrastructure of electricity markets also does not allow the participation of small players in the market due to insufficient incentives (Andoni, 2019). But all the challenges related to the energy market, energy demand, and energy supply can be easily overcome by replacing non-renewable energy sources with renewables, i.e. energy transition towards green energy. BCT provides flexibility, inclusiveness, and incentives in the field of energy markets, paving the way for energy transition in the years to come. Decentralized energy markets through BCT, contrary to traditional energy markets, would allow the producers and consumers to come together and meet their energy demand rationally, in a way that is both economical and sustainable. This decentralization of energy markets, also termed the "Energy Internet" facilitates the interconnection and energy coupling from different areas and sources. BCT facilitates the energy internet by enabling transparent and safe energy distributions involving both sellers and buyers. Many companies have pitched into the energy markets and explored the benefits of DLT through BCT to achieve energy sufficiency without emitting harmful gases. The growth of this technology cannot be witnessed in a year or two but would require a long gestation period. The World

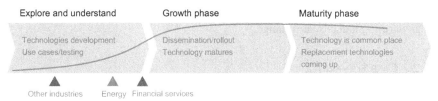

Growth phase Maturity phase

Technologies development   Dissemination/rollout   Technology is common place
Use cases/testing   Technology matures   Replacement technologies
                                                    coming up

Other industries   Energy Financial services

FIGURE 5.3 The growth of BCT in the energy markets as classified into three different phases by the the World Energy Council. Source: Author.

Energy Council has therefore classified the growth of BCT in energy markets into three different phases as given in Figure 5.3.

In Phase 1, **Exploration Phase**: The companies and the economies willing to step ahead for energy transition initiate pilot projects and startups to integrate BCT with existing energy infrastructure to increase the efficiency and reduce the cost together.

In Phase 2, **Growth Phase**: Significant progress is witnessed in the dissemination of technology across the countries. This technology initially developed to support cryptocurrencies like Bitcoin, has matured and has found its applications in energy sector. Blockchain has made substantial strides is in meeting the energy demand in an economically and environmentally friendly manner.

Phase 3, **Maturity Phase**: Blockchain technology has evolved beyond its initial applications and has gained maturity. Widely adopted in the energy sector, BCT becomes a common place and is ready for further advancements to meet the evolving needs of the industry.

It is also evident that most of the developed and developing countries like India, China, France, and many other countries have embarked on the journey of energy transition already and are moving towards phase 2, i.e. growth phase.

5.5 APPLICATION OF BLOCKCHAIN TECHNOLOGY

Some real cases of energy transition through the use of BCT in different countries have been discussed in this section.

5.5.1 CASE SCENARIO 1

A firm named Powergrid has developed BCT for trading energy. Households and businesses using solar energy are easily allowed to sell their excess energy generated to people nearby and encourage the adoption of renewables. This has not only helped people to lower the cost of energy but also helped to combat the threat of carbon emissions by minimizing the dependence on coal and oil. BCT has enabled resilience and flexibility to the electricity grids, allowing the consumers to track and trace every unit of electricity generated and consumed. BCT has not only helped with consumption but the decentralization of grids has enabled people to participate in energy markets, resulting in generation of income and improvement of lives. BCT is viewed as one of the effective solutions to create a stable and efficient power system in the present and future where the global economies are fighting against global warming and climate change.

5.5.2 CASE SCENARIO 2

5.5.2.1 Tata Power—DDL, Delhi

The BCT offered by Power Ledger's technology has helped the company to trade 2MW of solar power of 2MW in Delhi, India, serving 9.5 million customers. The company has been successfully selling power/excess energy to end consumers and local businesses at an effective price. This technology-driven energy trading has fostered the adoption of solar energy, which has also facilitated the country in overcoming issues like power deficit, climate change, etc.

5.5.3 CASE SCENARIO 3

5.5.3.1 TDED, Thailand

Peer-to-peer energy trading has been developed in Thailand with the help of BCT. The company has successfully traded around 112 MW of solar energy every month since 2022. TDED has dealt with energy problems differently by educating people and changing their behavior towards the adoption of solar technologies, enabling the new journey of energy transactions effectively. TDED has successfully created a sense of optimism by generating and trading 15 MW of solar energy in Chiang Mai, a city of Thailand. The pilot projects have been completed successfully, and the country is gearing up for its energy transition in the next few years.

5.5.4 CASE STUDY 4

5.5.4.1 Ekwateur, France

A firm in France, Ekwateur, has grown rapidly with more than 300,000 consumers today. The firm, with the help of BCT, accounts the units and sources of energy consumed. The consumers are able to know the type of energy consumed, and BCT in turn enables the consumers to customize their energy mix according to price and source. The company certifies and settles the energy transactions every 30 minutes, and consumers can choose the cheapest energy that could be either from a rooftop next to their door or from a wind farm close to their residence.

5.5.5 CASE STUDY 5

5.5.5.1 Enerchain, Germany

Enerchain, a project of Germany launched in 2017, has developed a technical infrastructure facilitating wholesale trading of energy, especially power and gas, in a decentralized manner using BCT by eliminating the intermediaries. This is the first project to enable spot and forward contracts in both power and gas markets, which has led to fall in the operational cost to a great extent.

All the cases discussed above reflect the fact that BCT has driven the energy revolution across the globe by decreasing our dependence on fossil fuels like coal and oil and successfully meeting our energy requirement from renewables.

5.6   ADVANTAGES OF BCT IN ENERGY SECTOR

Energy sector with blockchain technology provides the stakeholders with various advantages like reduced cost, transparency, flexibility, decentralization, etc. A comparison between the current energy market and the energy market with blockchain technology (Global P., 2017) given in Figure 5.4 would highlight the advantages of BCT in the energy sector.

5.6.1   DECENTRALIZATION AND TRANSPARENCY

Blockchain technology operates on a decentralized network that ensures control by multiple entities over the entire system. The degree of transparency also increases and, in turn fosters trust among the participants. Real-time monitoring of energy transactions becomes easy, effective, and efficient. BCT enables secure and real-time updates of energy use through an immutable ledger, ensuring greater efficiency and control over energy sources. Since the data cannot be altered, it improves regulatory compliance across the board and establishes a high degree of transparency.

5.6.2   COST REDUCTION

Introduction of BCT in the energy sector can help the energy sector cut down the expenses by eliminating inefficiencies, along with high degree of transparency and security. Several stages are involved in the production and distribution of energy, and all the parties involved in these stages can communicate efficiently through BCT, therefore helping in attaining optimization and cost reduction. Since BCT ensures an effective decentralized system, the role of management, market control, supervision, support, IT, etc., are very limited, which results in reduction in the cost of production and distribution.

5.6.3   DATA INTEGRITY AND SECURITY

In blockchain technology, the data related to energy transactions and usage is cryptographically secured and distributed through multiple nodes. The data cannot be altered or manipulated and would enhance data security and reduce the risk of fraud to a great extent.

5.6.4   SMART CONTRACTS

Automated processes such as meter readings, billing, and payments can help energy companies to reduce overhead costs, which would result in increased efficiency in the operating process.

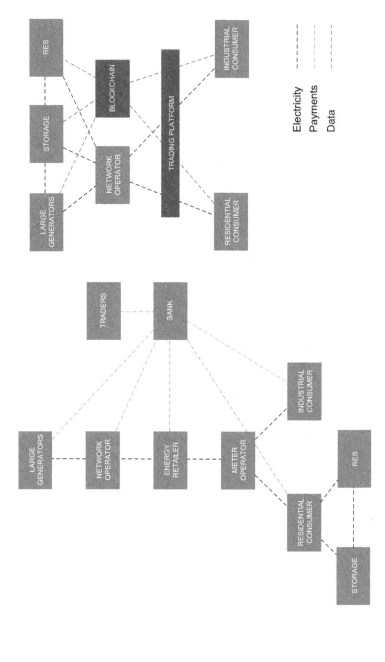

FIGURE 5.4 The advantages of BCT in the energy sector. Source: Author.

5.6.5 Energy Trading and Peer-to-peer Transactions

Energy trading between producer and consumer without any intermediaries can lead to efficient utilization of renewable energy sources. Currently, energy trading systems record the transactions in the conventional database, which is mutable and not transparent. Blockchain technology integrated with IoT and AI enables energy producers to sell the power generated directly to the consumers, thereby helping to balance the demand and supply of energy very efficiently.

5.6.6 Traceability and Certification

The consumers across the globe are concerned about the environment and therefore adopt green and clean energy in their day-to-day life. Blockchain technology provides the information related to the origin and life cycle of renewable energy through an immutable ledger. Transparent information provided to the consumers enhances the credibility of renewable energy projects and helps to build trust among the consumers. Transparency, details of its origin, and immutability can build trust among consumers and can result in behavioral change among users (Ahl, et al., 2019). It can therefore be opined that BCT in the energy sector could be useful for the economies working towards energy transition and achieving net-zero targets. The above-mentioned advantages of BCT underscore the fact that it is the only solution to address the present-day challenges of the energy sector.

5.7 CHALLENGES OF BCT IN ENERGY SECTOR

BCT can help the decentralization of the energy sector and optimize energy use to a great extent. Transparency and immutability could help create a positive change in the behavior of energy players, resulting in economic efficiency and environmental protection. However, there are many challenges that come in the way of the adoption of BCT in the energy sector and this section of the chapter highlights the key challenges of this technology in the area of energy trading.

Firstly, the integration of BCT with the currently existing technology is the greatest challenge as it would require huge investments.

Secondly, measures have been taken to reduce the cost of energy supply and to promote environment-friendly technologies to overcome energy-related issues, and BCT is opined as one such technology to accelerate the renewable energy usage. However, amendments in the current regulatory framework would also require a significant time period, which would delay the decentralization process.

Thirdly, a lack of flexibility and standardization would slow down the adoption of blockchain technology in the energy sector, as any further changes in the system have to be approved by the system nodes, which may lead to mistrust and fragmentation (Andoni, 2019).

Despite various challenges associated with BCT, it is a well-known fact that reliable and affordable energy sources are needed to foster the growth of an economy, and it is possible through the adoption of this technology.

5.8 DISCUSSION

Blockchain technology can go a long way in creating transparency and accuracy in terms of decentralized energy supplies. Financial incentives and energy policies have developed renewable energy sources, but the transition can be possible through the application of BCT. Deployment of smart meters in developed countries has helped nations reduce the greenhouse gas emissions and reach the net-zero target well in advance. Fifty-three million smart meters for electricity supply have been planned in the UK, which in turn requires a huge investment, approximately $2 trillion per year for the USA. A report published by the UK stated the capacity of BCT not only to revolutionize financial markets, as in the case of Bitcoins and Cryptocurrency, but in the energy markets by enabling transparency, security, and accountability. BCT, believed to be a game-changer for energy suppliers can help achieve the low-carbon transition.

There is a huge scope for BCT in the process of energy transition, but the scalability, cost, and security are considered major challenges in the application of BCT in the energy sector. Various micro and macro aspects today require the attention of policymakers to regulate the layout to accelerate BCT and its adoption in energy markets. Along with this, the policymakers of the nations can frame effective strategies to cope with the challenges of BCT in the energy sector and can strengthen the energy supply without compromising the environmental challenges.

5.9 CONCLUSION

The chapter has provided an in-depth discussion about BCT in the energy sector as it is considered a disruption. The energy sector across the globe is undergoing serious challenges related to transparency, safety, security, environmental harm, etc., and BCT can provide a long-run sustainable solution through an effective decentralization system.

The global market for BCT is said to grow between 50% and 75% annually, reflecting rapid digital transformation, and the industries have successfully raised $23.7 billion since 2013. Decentralized power generation and consumption through BCT, contrary to the centralized energy markets, enable a dynamic balance between the demand and supply of energy across the electrical infrastructure very efficiently and effectively.

Many countries across the globe have successfully initiated BCT in the energy sector in the last few years, but commercially successful business in the energy blockchain space is yet to be witnessed. BCT enables players to manage demand in an effective manner, but the challenges are primarily found in the deployment of blockchain applications for EV infrastructure. Technology acceptance and public adoption limit the use of BCT in the energy sector, which could be overcome through educating people about the economic and environmental benefits arising from the application of BCT in energy markets.

REFERENCES

Andoni, M., Robu, V., Flynn, D., Abram, S., Geach, D., Jenkins, D., & Peacock, A. (2019). Blockchain technology in the energy sector: A systematic review of challenges and opportunities. *Renewable and Sustainable Energy Reviews, 100*, 143–174.

Ahl, A., Yarime, M., Tanaka, K., & Sagawa, D. (2019). Review of blockchain-based distributed energy: Implications for institutional development. *Renewable and Sustainable Energy Reviews, 107*, 200–211. https://doi.org/10.1016/j.rser.2019.03.002

Ahl, A., Yarime, M., Goto, M., Chopra, S., Kumar, N., Tanaka, K., & Sagawa, D. (2019). Exploring blockchain for the energy transition: Opportunities and challenges based on a case study in Japan. *Renewable and Sustainable Energy Reviews, 117*, 109488. https://doi.org/10.1016/j.rser.2019.109488

Antonopoulos, A.M. (2014). *Mastering Bitcoin: Unlocking Digital Cryptocurrencies*. O'Reilly Media, Inc.: Sebastopol, CA, USA.

Al-Jaroodi, J., Mohamed, N. (2019). Blockchain in industries: A survey. *IEEE Access, 7*, 36500–36515.

Andoni, M., Robu, V., Flynn, D., Abram, S., Geach, D., Jenkins, D., McCallum, P., & Peacock, A. (2019). Blockchain technology in the energy sector: A systematic review of challenges and opportunities. *Renewable and Sustainable Energy Reviews, 100*, 143–174.

Burger, C., Kuhlmann, A., Richard, P., & Weinmann, J. (2016). Blockchain in the energy transition a survey among decision-makers in the German energy industry. https://shop.dena.de/fileadmin/denashop/media/Downloads_Dateien/esd/9165_Blockchain_in_der_Energiewende_englisch.pdf, (accessed 15 May 2017)

BlockchainIn Government: Welfare, Foreign Aid & More Uses | Built In. https://builtin.com/blockchain/blockchain-in-government (accessed 27 August 2022).

Chen, H.S., Jarrell, J.T., Carpenter, K.A., Cohen, D.S., & Huang, X. (2019). Blockchain in healthcare: A patient-centered model. *Biomedical Journal of Scientific & Technical Research, 20*, 15017–15022. https://pubmed.ncbi.nlm.nih.gov/31565696/ (accessed 27 August 2022).

Chen, Y., Cheng, L., Lee, C.-C., & Wang, C. (2021). The impact of regional banks on environmental pollution: Evidence from China's City commercial banks. *Energy Economics, 102*, 105492. https://doi.org/10.1016/j.ene co.2021.105492.

Chouhan, V., Goswami, S., Dadhich, M., Saraswat, P., & Shakdwipee, P. (2021). Emerging opportunities for the application of blockchain for energy efficiency. *Blockchain 3.0 for Sustainable Development, 10*, 63.

Grant, D., Zelinka, D., & Mitova, S. (2021). Reducing CO_2 emissions by targeting the world's hyper-polluting power plants*. *Environmental Research Letters, 16*, 094022. https://doi.org/10.1088/1748-9326/ac13f1

Gerlak, A., Weston, J., McMahan, B., Murray, R.L., & Mills-Novoa, M. (2018). Climate risk management and the electricity sector *Clim. Risk Manage, 19*, 12–22.

Global, P. (2017). Blockchain-an opportunity for energy producers and consumers. *PwC Global Power and Utilities*.

Grinberg, R. (2011). Bitcoin: An innovative alternative digital currency. *Hastings Science & Technology Law Journal, 4*, 159.

International Energy Agency. (2009a). *Sectoral Approaches in Electricity: Building Bridges to a Safe Climate*. Paris: International Energy Agency.

International Energy Agency. (2009b). *How the Energy Sector Can Deliver on a Climate Agreement in Copenhagen*. Paris: International Energy Agency.

International Energy Agency. (2021). "Net Zero by 2050: A Roadmap for the Global Energy Sector" International Energy Agency, www.iea.org

Jie, H., Khan, I., Alharthi, M., Zafar, M.W., & Saeed, A. (2023). Sustainable energy policy, socio-economic development, and ecological footprint: The economic significance of natural resources, population growth, and industrial development. *Utilities Policy, 81*, 101490, ISSN 0957–1787, https://doi.org/10.1016/j.jup.2023.101490

Khan, I., Zakari, A., Zhang, J., Dagar, V., & Singh, S. (2022). A study of trilemma energy balance, clean energy transitions, and economic expansion in the midst of environmental sustainability: New insights from three trilemma leadership. *Energy, 248*, 1–11. https://doi.org/10.1016/j.energy.2022.123619

Lee, C.-C., Xing, W., & Lee, C.-C. (2022). The impact of energy security on income inequality: The key role of economic development. *Energy, 248*, 123564. https://doi.org/10.1016/j.energy.2022.123564

Liu, H., Mansoor, M., Al-Faryan, M.A.S., Khan, I., & Wasif, M. (2022). Impact of governance and globalization on natural resources volatility: The role of financial development in the Middle East North Africa countries, Resour. *Pol, 78*, Article 102881. https://doi.org/10.1016/j.resourpol.2022.102881

Mylrea, M., & Gourisetti, S. N. G. (2017, September). Blockchain for smart grid resilience: Exchanging distributed energy at speed, scale and security. In *2017 Resilience Week (RWS)* (pp. 18–23). IEEE.

Office of Gas and Electricity Markets (Ofgem). (2017). Transition to smart meters, https://www.ofgem.gov.uk/gas/retail-market/metering/transition-smart-meters, (accessed 6 Dec 2017).

Oum Kumari, R., Sunita, C., & Sani, M. (2019). Environmental Kuznets curve for sustainable development. *International Journal of Innovative Technology and Exploring Engineering (IJITEE)*, 8, 12S.

Press AXONI. https://axoni.com/press/axoni-distributed-ledger-network-for-equity-swap-processing-goes-live-with-leading-market-participants/ (accessed on 27 August 2022)

Shrier, D., Wu, W., Pentland, A. (2016). *Blockchain& Infrastructure (Identity, Data Security)*. Cambridge, MA, USA: Massachusetts Institute of Technology.

Strachan, M. (2022). Santander Launches the First Blockchain-Based International Money Transfer Service across Four Countries. p. 2. https://www.santander.co.uk/about-santander/media-centre/press-releases/santander-launches-the-first-blockchain-based-international-money-transfer-service-across-four (accessed 27 August 2022).

Taghizadeh-Hesary, F., Zakari, A., Yoshino, N., & Khan, I. (2022). "Leveraging on Energy Security to Alleviate Poverty in Asian Economies. *The Singapore Economic Review*. https://doi.org/10.1142/s0217590822440015

Tyan, I., Yague, M., & Guevara-Plaza, A. (2020). Blockchain technology for smart tourism destinations. *Sustainability, 12*, 9715.

UNEP. (2022). https://www.unep.org/resources/emissions-gap-report-2022

Wen, H., Lee, C.-C., & Zhou, F. (2022). How does fiscal policy uncertainty affect corporate innovation investment? Evidence from China's new energy industry. *Energy Economics*, 105, 105767. https://doi.org/10.1016/j.eneco.2021.105767

Walport, M. Distributed ledger technology: Beyond blockchain. https://assets.publishing.service.gov.uk/government/uploads/system/uploads/attachment_data/file/492972/gs-16-1-distributed-ledger-technology.pdf

Zhou, S., & Brown, M. A. (2017). Smart meter deployment in Europe: A comparative case study on the impacts of national policy schemes. *Journal of Cleaner Production*, 144, 22–32. https://doi.org/10.1016/j.jclepro.2016.12.031

6 An Overview of Digital Signature Law and Practice Adopted in India

Manickavasagam, V. Vijaya, and Bharathi Ravi

6.1 INTRODUCTION

Cryptography, which translates to "secret writing," was first the theory of encrypting data or communications. This science was developed thousands of years ago and is today more significant than ever. Without cryptography, life is unimaginable, particularly internet communication. Two parties can safely communicate across a secure connection because of cryptography. These days, cryptography has many uses and is much more than just message encryption – creating and designing new secure cryptographic primitives, like cryptographic hash functions, identity-based encryption methods, and digital signature techniques, among many other cryptographic building blocks. It also focusses on enhancing these cryptographic primitives' efficiency and security proofs. In today's rapidly expanding digital world, online authentication has become an essential tool for any business. Signatures play an important role in commerce as well as in law where daily transactions take place. It is a symbolic representation of an individual and a portrayal of the intent in it. A signature attests to the legality of a specific transaction. Digitalisation being the order of the day has transformed the landscape of business. Handwritten signatures are insufficient for internet contracts because they can easily be manipulated and forged in online transactions. The rapid development of electronic or internet-based business has given rise to a new paradigm for corporate operations. Detailed guidelines are given to ensure the information's legitimacy in this extensive transmission line system. Stamp-based digital signature technology is adopted in this special place, and it has data security in the field of computer systems and technology (Mason, 2006). However, despite the fact that online businesses have a strong interest in information security, the technology is trying to be a great fit to meet the security requirements of digital signature seals (Panko 2004). There are many types of electronic signatures that can be used for authentication, while digital signatures are one of the most effective tools. Digital signatures provide a workable solution that can help to bridge the gap between going completely paperless and doing away with the need to print documents for signatures. It is a legally enforceable electronic tool. Digital signatures permit the fast, affordable digital substitution in place of slow and expensive paper-based approval

DOI: 10.1201/9781003453109-6

processes. Authentication, repudiation, and verification of electronic data are integral part of any electronic transactions. Unless these procedures are undergone, the authentication and security of these transactions remain virtual. The digital signature process is used to ensure the security and authentication of electronic data. A digital signature is a mechanism to authenticate data, meaning that it confirms that the document you have received is from the person who has claimed to be the sender and that its content hasn't been changed since they made it. Similar to how traditional systems use stamps, seals, or signatures to establish the authenticity of paper documents, digital signatures authenticate the electronic record. By adding this digital signature, a subscriber of the digital signature wishes to authenticate any electronic record, and it establishes the authenticity of that record. In today's digital world, on an average trillions of transactions are enabled. Among the most pressing and unsolvable outstanding issues in cryptography are effective digital signatures and information-theoretic security that guarantees the integrity, authenticity, and non-repudiation of data. Now it is used for electronically signing contracts between two countries, papers, filing taxes, and more.

UNCITRAL Model Law on Electronic Commerce established uniform standards for digital signature laws for e-business and e-commerce, which was released by the UN in 1996. The majority of the countries began to adopt it. When a document is digitally signed, it has legal bounds. Laws pertaining to digital signatures may differ slightly among nations.

Countries also differ in terms of the awareness of and proportion of people utilising digital signatures. When compared to developed countries, the percentage of internet users in emerging countries is significantly lower, and this is directly correlated with the number of users of digital signatures. In knowing about the digital signature in different countries, and when and where the digital signature is used a study is carried out by diving the countries as developed and merging economies. Along with it the use of digital signature in KYC form of banking is also dealt.

6.2   HISTORY OF DIGITAL SIGNATURE

In 1976, Diffie and Hellman laid the groundwork for public key encryption and introduced the notion of digital signature. Shortly after in 1977, Rivest, Shamir, and Adleman developed the RSA encryption method, which allowed for meaningful homomorphic operations on signatures and could also be used to create crude digital signatures, but these were far from safe. Soon later, in 1979, Lamport created a digital signature system that, because it could only be used once per key pair, satisfied a relatively weak security feature (Shafi Goldwasser, Silvio Micali and Ronald L. Rivest, 1988). Although this might not appear essential at first, it is often easy to expand this method into a multi-use digital signature technique. One of the weakest presumptions in cryptography is the presence of one-way functions, upon which the construction is predicated. Additional initial constructs were provided and the first security specifications for digital signature methods were established by Goldwasser in 1988. They introduced a scheme called GMR signatures, which was demonstrably secure in every way. It might be argued that even if there has been significant advancement in creating (very) secure digital signature schemes, there is always room for improvement. One of the main areas of research is the creation of novel schemes that meet strict security requirements while being economical with regard to key and signature sizes and verification computation times.

The art of sending data in a way that a third party cannot interpret is known as cryptography. Any form can be used for the data. Encryption and decryption are the two fundamental procedures of cryptography (Gary C Kessler, 1998).

- Encryption is the process by which the sender transforms the original data into unreadable cipher data using a key, or, to put it another way, a set of rules.
- On the receiving end, decryption is the process of reversing the ciphertext into the original data using a key, or more precisely, a set of rules.

These are the two steps by which cryptography saves the information from sharing it with other than the desired person while transferring it. Though there are many types of cryptography, encrypt and decrypt cryptography can be classified into symmetric cryptography, otherwise known as secret key cryptography (SKC), asymmetric cryptography, or public key cryptography (PKC) and hash function.

Digital Signatures' Security Considerations, Correctness and Soundness of a digital signature technique are primary requirements in Security Notions of Digital Signatures. Informally, correctness is defined as "the scheme works," that is, every signature for every message created by the scheme's signing algorithm using any secret key. It should always result in a valid output from the verification algorithm when the signature, message, and associated public key are entered. In layman's terms, "the scheme is secure" indicates that no effective algorithm or adversary can break any of the scheme's security properties with a non-negligible probability. In theory, it would be preferable if it were impossible to create a legitimate message signature without the secret key.

GMR88 originally introduced several key security concepts. An attack result, which outlines the objective an adversary must accomplish in order to breach the scheme's security, and an attack model, which characterises the capability and strength of an adversary, combine to formalise the idea of security. There are many security notions depending on the model and the goals

Existentially Unforgettable under Non-adaptive Chosen-message Attacks (EUF-naCMA): An adversary must select a set of messages prior to viewing the scheme's public key. The public key and matching signatures for the selected messages are then given to the attacker. The opponent wants to create a new, random message of his choosing and forge a legitimate signature on it.

Existentially Unforgettable under Chosen-Message Attack (EUF-CMA): After obtaining the public key, the adversary can choose messages and get the accompanying signatures in an adaptive manner, whereas adaptively is defined as relying on the public key or prior signatures. Once more, the adversary wants to forge a legitimate signature on a fresh, random message of his choosing.

6.3   THE THEORETICAL BACKGROUND OF DIGITAL SIGNATURE

The following are the benefits and uses of digital signature.

6.3.1   SECURITY

Digital signatures include security features to guarantee that a legal document hasn't been altered and that signatures are genuine. Asymmetric cryptography, PINs, checksums, and cyclic redundancy checks (CRCs), as well as CA and trust service provider (TSP) validation, are security features.

6.3.2   TIMESTAMPING

This offers the date and time of a digital signature and is helpful when timeliness is important, such as for stock deals, the printing of lottery tickets, and legal processes.

6.3.3   GLOBALLY ACCEPTED AND LEGALLY COMPLIANT

The vendor-generated keys are created and stored securely according to the public key infrastructure (PKI) standard. As digital signatures become the norm globally, more nations are recognising their legal significance.

6.3.4   SAVINGS IN TIME

Digital signatures make it easier for businesses to obtain and sign documents rapidly by streamlining the laborious processes involved in traditional document signing, storage, and exchange.

6.3.5   COST SAVINGS

By becoming paperless, businesses can save money on the materials, labour, employees, and office space traditionally required to store, handle, and transfer papers.

6.3.6   BENEFITS FOR THE ENVIRONMENT

Reducing paper use also lessens the physical waste paper produces and the harmful effects that moving paper documents have on the environment.

6.3.7   TRACEABILITY

With the use of digital signatures, firms can maintain internal records more easily. There are fewer chances for a manual signer or record keeper to make a mistake or lose something when everything is recorded and stored digitally. The usage of digital signature is discussed in the next paragraph.

Generally, digital signature tools and services are used contract heavy industries such as government for public and private laws, and budgets. Digital signatures are used by governments all over the world to process tax returns, confirm business-to-government transactions, ratify laws, and manage contracts. When utilising digital signatures, the majority of governmental organisations are required to abide by tight laws, rules, and guidelines. Smart cards are also widely used by businesses and governments to identify their workers and constituents. These are physical cards with a chip inside that holds a digital signature and grants the cardholder access to the systems or actual buildings of an institution. In the healthcare sector, digital signatures are used to streamline administrative and therapeutic procedures, boost data security, enable e-prescribing, and manage hospital admissions. Digital signatures are used by manufacturing organisations to streamline operations such as product design, quality control, production improvements, marketing, and sales. For contracts, paperless banking, loan processing, insurance documentation, and mortgages digital signatures are used. Digital signatures are used by Bitcoin and other cryptocurrencies to verify the blockchain. With digital assets, such as music, artwork, and films, digital signatures are utilised to secure and track these kinds of NFTs everywhere on the blockchain.

6.4 CLASSES AND TYPES OF DIGITAL SIGNATURE

There are three types of digital signatures:

Class 1: Since they are solely validated based on an email ID and username, this sort of DSC cannot be used for official business documents. Class 1 signatures offer a fundamental level of security and are applied in settings where there is little chance of data compromise.

Class 2: These DSCs are frequently used for electronic (or "e-")filing of tax papers, including as returns for both the goods and services tax and the income tax. Class 2 digital signatures check the identity of the signer against a database that has already been confirmed. In settings with moderate risks and effects of data compromise, class 2 digital signatures are utilised.

Class 3: The highest level of digital signatures, Class 3 signatures demand that individuals or organisations present themselves before a CA to establish their identity before signing. Class 3 digital signatures are used in e-auctions, e-tendering, e-ticketing, court filings, and other settings where there are significant risks to data or repercussions from a security breach.

6.5 THE USE OF PUBLIC KEY INFRASTRUCTURE (PKI) AND PRETTY GOOD PRIVACY (PGP) WITH DIGITAL SIGNATURE

Public key infrastructure (PKI) and the pretty good privacy (PGP) encryption programme are both used in digital signatures to minimise any potential security risks associated with transmitting public keys. They authenticate the sender's identity and confirm that the sender's public key indeed belonged to that person.

The digital certificate of the executives (create, disperse, approve) is the key to the PKl framework. A digital certificate functions similar to an authentic ID card as an electronically recognised proof card. It is provided by a Certifying Authority – a verification officer. People can use it to connect with one another and understand each other's personalities. In a written document, marking is meant to confirm it. Additionally, it serves two purposes: first, it ensures the authenticity and genuineness of the document. Since it is hard to refuse their own signature, which validates the markings made on it; and second, it ensures that the document is legitimate and authentic because signature is hard to manipulate as deviated cryptography is used with digital signatures.

Public key infrastructure (PKI) is a framework for services that create, manage, distribute, and track public key certificates. Symmetric key and public key cryptography are used in PGP, a variant of the PKI standard, although it differs in how it links public keys to user identities. PKI uses a web of trust, whereas PGP uses a CA to validate and associate a user identity with a digital certificate. PGP users decide which identities to verify and who they trust. PKI users respect reputable CAs.

The strength of the private key security determines how secure a digital signature can be. Without PKI or PGP, it would be harder to identify someone or revoke a compromised key, and it would be simpler for malevolent actors to pass themselves off as legitimate individuals.

The difference between digital signature and e-signature.

Digital Signature	Electronic Signature
Not only uses algorithm but also a cryptographic process to validate the sequence of data, to verify the origin of the signature, and authenticity of a document.	Uses electronic sound symbols or processes attached to or associated with a contract or record to verify the origin of the signature.
Must be issued by a certificate authority (CA).	Can be any electronically applied signature.
It is a type of electronic signature.	Is a broader term that encompasses digital signature in its definition.
Provides cryptographic proof of the authenticity and integrity of a document and the signer's signature.	Confirms a signer's intent to sign a document but does not always provide proof of a signer's identity or the document's integrity.

Having provided the theoretical background, uses, and classes of digital signatures, the review of literature is discussed next.

6.6 REVIEW OF LITERATURE

According to Hart (1998):

> Literature review is a collection of available documents on relevant topics which may either published of unpublished, Literature review includes data, information, ideas and evidences which have taken from a definite viewpoint of the topic. The viewpoint

should have a certain aim and it should give the idea about how the topic will be investigated.

From a conventional standpoint, written signatures validate and approve paper documents, and they are a really good way to provide authenticity. A similar procedure is necessary for electronic documents. The requirement for validation and verification of electronic documents is met by digital signatures, which are only a string of ones and zeroes generated by applying a digital signature algorithm. Verification alludes to the process of confirming the document's sender, whereas approval refers to the means of assuring the document's content. There is no distinction made between the terms message and document in speaking about the features of digital and conventional signatures (Cabanellas 2018). The distinguishing characteristics of a customary signature are as follows: it is relatively easy to establish that the signature is bona fide; it is difficult to produce; it is not transferable; it is difficult to alter; and it cannot be repudiated, ensuring that the signatory cannot subsequently deny signing. Given that digital signatures are being used in sensitive but useful applications like secure email and credit card transactions over the internet, they should have all the features of a traditional signature listed above in addition to a few more.

Given that a digital signature consists just of zeroes and ones, it would be desirable for it to possess the following attributes: it must be reasonably easy to deliver; it must be generally easy to understand and verify the validity of the digital signature; it must be computationally impossible to fabricate a digital signature by creating a fake digital signature for a given message or by building another message for an existing digital signature; the signature must be a piece of design that depends on the message being marked (so, for the equivalent originator, the digital signature is diverse for various documents); and it (for the equivalent originator, the digital signature is diverse for various documents); it must be computationally impossible to manufacture a digital signature by creating a false digital signature for a given message or by building another message for an existing digital signature. The signature must make use of some data that is noteworthy to the sender in order to prevent both fabrication and disavowal. Additionally, copies of the digital signature must be useful in order to facilitate future discussions about its incapacity (Elisa). Several techniques, known as validation procedures, have been developed to ensure that the document received is unquestionably from the promised sender and that the content has not been altered. However, because confirmation systems are inadequate, message verification processes cannot be used directly as digital signatures (Duggal 2002). For example, message validation guarantees that the two groups are exchanging messages from an outsider, but it does not guarantee that the two groups are at odds with each other. Basic validation strategies result in signatures that are the message's length when expanded. Important concepts and language. The documents (messages/data) that need to be agreed upon, as well as some private information that is uniquely possessed by the sender, determine how digital signatures are processed. In reality, to obtain the message digest, a hash capacity is attached to the message rather than using the complete message. In this particular case, a hash operation generates a fixed-size message digest as output after accepting a self-assured approximated message as input (Daniel). *Meyer v. Uber* and *O'Connor v. Uber*," she stated. Since it was displayed on "a modest iPhone screen when most drivers are going to go on obligation," the offended party claimed that there was unquestionably no valid understanding in the main issue, and the court was asked to evaluate this argument. The court rejected that argument, ruling that as long as someone has the opportunity to read an agreement, it is unnecessary. Since it was displayed on "a modest iPhone screen when most drivers are going to go on obligation," the offended party claimed that there was unquestionably no valid understanding in the main issue, and the court was asked to evaluate this argument. The court rejected that argument, ruling that as long as someone has the opportunity to read an agreement, it is unnecessary.

Through relevant publications and papers, information is obtained concerning the theoretical underpinnings, historical development, and legal framework surrounding digital signatures in India as opposed to other industrialised nations. By perusing reliable official websites, one may learn about the applications for digital signatures as well as the prerequisites for obtaining one. Emails sent to the Controller of Certification Agencies (CCA) and reliable official websites provide information about the percentage of users utilising digital signatures. In order to bolster the study, a survey was conducted. By framing the survey question a clear literature with deep understanding of the study has been made with previous researches in the same area.

6.7   OBJECTIVES OF THE RESEARCH

1. To review the historical development of digital signature.
2. To identify and analyse awareness about digital signature in an emerging economy (country); India.
3. Analysis and apprehending the information about digital signature in legal perspective with respect to Income Tax of India.
4. To understand about the security of digital signature in India.
5. To do a comparison of digital signature with traditional signature.

6.8   RESEARCH QUESTIONS

RQ1. What are the key parameters of digital signature?
RQ2: What are the challenges in improving digital signature in developing countries?
RQ3: In what field is digital signature used in India?

It uses asymmetric cryptography to encrypt the data, providing reason to believe that the data was send by the claimed sender.

6.9   DIGITAL SIGNATURE LAWS

One popular method of document authentication is using electronic digital signatures. To our surprise, the law pertaining to digital signatures is old. Contracts and business transactions were conducted by telegraph machines at the beginning of the 19th century; the signer's identity was confirmed via a signature encrypted with Morse code. The digital signature law was created as a result. Now signing contracts and other legal papers even when the parties are on separate sides of the globe these technologies. As early as 1869, the Supreme Court of New Hampshire upheld the constitutionality of digital signatures. The law pertaining to digital signatures was updated as a result of the widespread usage of fax machines. The software licence for ATMs is then clicked to be accepted, and laws pertaining to digital signatures have been updated frequently to keep up with technological advancements. The legal framework around digital signatures has changed recently due to the rise of e-business and e-commerce. In order to provide universal standards for digital signature legislation for the worldwide e-business and e-commerce industry, the UN released the UNCITRAL Model Law on Electronic Commerce in 1996. A number of governments quickly began putting the digital signature law into effect. There have been published laws such as the Electronic Signature Directive of the European Union and the Electronic Signature Law 1999 of South Korea. In the year 2000 countries like Canada, the United Kingdom, the United States (Electronic Signatures in Global and National Commerce (E-SIGN) Act), and India enforced the law.

The UNCITRAL Model Law on Electronic Signatures was adopted in Vienna on July 5, 2001, in response to the growing use of electronic document circulation in international trade and relations. Its goals were to encourage further adoption of the electronic signature concept. Additionally, the document came to represent the assurance that specific electronic signature techniques could be applied to transactions that would have major legal ramifications. The European Union aimed to simplify and increase transparency of regulations concerning electronic digital signatures by enacting this statute.

This regulation defines an electronic signature as data in electronic form that is used to identify the signatory and is connected to or related with the exchange of information. The signature implies an automatic agreement with the contents of the document by the person who left it. It is vital to ascertain the precise provisions of this nation's legislative legislation regarding electronic digital (EDS) prior to finalising an electronic transaction with a natural or legal entity situated abroad. The solicitors putting the document up for signature take on this duty. Nowadays, just fifty or so nations frequently employ electronic signatures.

There are three types of laws considered worldwide. They are discussed below.

6.9.1 Prescriptive Laws

There are very few prescriptive laws. These standards specify the kinds of signature technologies that are permissible and demand a particular technical approach for electronically signing the document. Prescriptive e-Signature regulations are found only in a selected nation, namely Brazil, India, Israel, and Malaysia.

6.9.2 Minimalist Laws

As the name suggest, these regulations permit e-Signatures to be widely enforced with little limitations. These laws are not technology-specific and allow the use of e-Signatures for all users. Like a shield, minimalist rules offer the broadest protection and are effective in the majority of circumstances. These rules are typical in nations including Canada, Australia, New Zealand, and the United States.

6.9.3 Two-tier Laws

Two-tier rules give digital signatures more weight as evidence and allow the use of e-Signatures. This is a combination of prescriptive and minimalist legislation. Similar to minimalist laws, two-tier laws establish a class of authorised technologies, similar to prescriptive laws, but they also accept all or most e-Signatures on a technology-neutral basis. Two-tier e-Signature rules have been implemented by China, South Korea, and most European countries.

6.10 UNDERSTANDING DIGITAL SIGNATURE LAW AROUND THE WORLD

6.10.1 United States of America

The Uniform Electronic Transactions Act (UETA), which established a legal framework for the use of e-Signatures, was created in 1999 by the Uniform Law Commission. The U.S. Congress passed the Electronic Signatures in Global and National Commerce Act in 2000 to ensure the legality and validity of contracts entered electronically, thereby facilitating the use of electronic records and e-Signatures in interstate and international commerce. All business agreements, financial contracts are done by digital signature. Today the electronic signature is equal to writing. Irrespective of whatever may be the documents, it has its full legal force. Here even non-state certified businesses are issuing EDS. There aren't any certification centres as a result. The signature classification is also absent, with the exception of the highly responsible sectors of real estate, healthcare, and government contracts. The necessity of executing an agreement is one distinctive feature of this area of US law. This implies that companies need to write a contract outlining their shared commitment to conducting business and supplying basic paperwork.

6.10.2 United Kingdom

The UK belongs to the first group of nations that use EDS. This implies that electronic signatures have equal legal weight and full legal force as a written one. A certificate of this kind will be just as valid in court as a comparable written document. Small- and medium-sized enterprises are especially encouraged to adopt EDS since it enables them to transact with foreign partners without having to arrange in-person meetings. Additionally, consent to use electronic material is being requested in advance from representatives of big businesses.

6.10.3 CANADA

In Canada, electronic workflow and signatures are widely used in court procedures as well as private company and government contracts. Since electronic paperwork saves a great deal of time and money, the state not only distributes but even encourages the use of electronic signatures. It is far simpler to issue a personal signature using a hidden key.

6.10.4 AUSTRALIA

e-Signatures are common and are regularly used for business transactions. Each State and Territory in Australia has its own electronic transactions legislation, which often reflect the Commonwealth ET Act, but include some specific exceptions. The law does not require data related to e-Signatures to be stored within the continent exclusively. Approximately 33% of people in this nation utilise EDS. The format of an electronic signature is not required. The present Electronic Transactions Act is in effect throughout Australia's Union and in any situation where a person's signature is necessary. The use of this technique of document certification requires the express approval of the individual. In practical terms, restrictions apply when the individual who obtained the electronic signature fails to verify that it complies with the required identity standards and when a witness is needed to sign the document. Furthermore, if a law or an act of a state entity mandates a handwritten signature, EDS will not be acknowledged.

6.10.5 BELGIUM

Digital signature is very common in Belgium. However they are subject to exceptions. Digital signatures based on certificates are used in new ways. Since no one may be forced to sign a contract electronically by law, there must always be the choice to sign with a handwritten signature. "A set of electronic data attributable to a specific person and demonstrating the preservation of the integrity of the content of the document can be regarded as a legally valid signature," according to the July 21, 2016, Act. This Act also known as the Digital Act is the most significant statute to date.

6.10.6 GERMANY

Germany, an EU member state, has been able to legally recognise e-Signatures since 2001 according to the German Signature Law, which was created following the approval of the EU Directive in 1999. In 2014, the EU Regulation No. 910/2014, popularly referred to as the eIDAS Regulation, took the role of the EU Directive. The Vertrauensdienstegesetz (VDG), often known as the German Trust Services Act, is a major piece of legislation in Germany that governs the use of e-Signatures. It facilitates the use of electronic trust services in accordance with the eIDAS Regulation. The German Civil Code, or BürgerlichesGesetzbuch (BGB), establishes when written form can be substituted with electronic form, among other things. In Germany, e-Signatures are not widely used. Nonetheless, the corporate community is using them more and more.

6.10.7 INDIA

The Information Technology Act of 2000 (IT Act) governs digital signatures that are based on certificates and electronic signatures. According to this regulation, both are equivalent to handwritten signatures. They are the recommended method for completing several government operations, like goods and service tax filings and electronic filing with the Ministry of Corporate Affairs. An electronic authentication method or process listed in the IT Act's Second Schedule must be included in valid e-Signatures.

6.10.8 INDONESIA

Ever since the Indonesian e-Signature Regulations were introduced, e-Signatures have become widely used. The following laws are regulations: Government Regulation 71 of 2019 on the Application of Electronic Systems and Transactions (GR71/2019); Law 11 of 2008 on Electronic Information and Transactions, amended by Law 19 of 2016; Minister of Communications and Informatics (MoCI) Regulation No. 11 of 2018 on Administration of Electronic Certification ("MoCI Regulation 11 of 2018"). Both certified e-Signatures, also referred to as digital signatures, and uncertified e-Signatures, which do not require a registered digital certificate provider, are accepted under these requirements.

6.10.9 SOUTH AFRICA

Although e-Signatures are frequently used, verbal contracts are still prevalent and do not always need to be written down or signed. The Electronic Communications and Transactions Act 25 of 2002's Section 13 serves as the primary legal framework for digital signatures. It is important to remember that the ECTA's Accreditation Regulations control the certification of authentication services and products that facilitate enhanced e-Signatures. A governing body may accept electronic licences, permits, payments, and documents under Sections 27 and 28 of the ECTA. At present there no laws that forbids the storage and processing of South Africa. However, the Protection and Personal Information Act of 2013 (POPIA) should be considered beyond South Africa for process and transfer.

6.11   DIGITAL SIGNATURE AND ITS VARIANTS

6.11.1   Dital Signature on Files

This file is kept on the computer. The file is used with a password in order to authenticate. The file is available for purchase from the provider or download from the internet bank. This file can be copied to another computer. The file is also compatible with USB sticks.

6.11.2   Digital Signature on Cards

This kind of digital signature uses a chip card to hold the digital signature. It is more expensive than the previous one. With this kind of authentication, a card reader that is USB, connected to the computer, is utilised for authentication. The company that provides the digital signature certificate (DSC) also provides the card reader. In addition, the card comes with the owner's portrait on it, which doubles as an identity card.

6.11.3   Digital Signature on Mobiles

There is e-authentication accessible for tablets and smartphones. This feature is provided by mobile bank ID. The application must be downloaded to the device and linked to the internet bank in order to utilise this feature. The Information Technology Act of 2000 made e-authentication lawful in India. Digital signatures are accorded the same legal weight as handwritten signatures, and electronically signed documents have the same legal standing as conventional papers. The Information Technology Act, which is based on an asymmetric cryptosystem, gives digital signatures the necessary legal validity.

6.12   E-AUTHENTICATION GUIDELINES FOR E-SIGN AND ONLINE ELECTRONIC SIGNATURE SERVICE (ISSUED UNDER ELECTRONIC SIGNATURE OR ELECTRONIC AUTHENTICATION TECHNIQUE AND PROCEDURE RULES 2015) LAST UPDATED 27TH JANUARY 2021

6.12.1   Explanation to Certain Terminologies

"e-Sign" or "e-Sign Service" is an online electronic signature service that, upon successful individual authentication through e-KYC services, facilitates instantaneous key pair generation, public key certification by the CA, and digital signature creation for electronic documents all within a single online service. "e-Sign user or e-KYC user or user or subscriber" is an individual requesting for e-Sign online electronic signature service of an e-Sign service provider. "e-KYC" means the transfer of digitally signed demographic data such as name, address, date of birth, Gender, Mobile Number, Email address, photograph etc of an individual collected and verified by e-KYC provider on successful authentication of same individual. "Response code" is the authentication number maintained by e-KYC provider to identify the authentication.

The digital signature certificates (DSCs) are granted by certifying authorities (CA) in accordance with the Information Technology Act, 2000, and rules made thereunder, following a successful verification of the applicant's identity and address credentials. To start with, the Second Schedule of the Information Technology Act of 2000 mentions e-KYC as an e-authentication service that CAs are supposed to run under these standards. CA may employ the same human resources and physical infrastructure for e-authentication. The level of security required for this service should match the level of security that the CA already upholds. Additionally, the audit of CA facilities must include the audit of e-authentication. The term "e-Sign Service Provider" (ESP) refers to the Trusted Third Party e-Sign-Online Electronic Signature Service of California.

6.12.2   ESP Requirements

6.12.2.1   2.0 e-KYC Providers

The applicable e-KYC services provider for e-Sign are:

1. UIDAI (online Aadhaar e-KYC services).
2. e-Sign User Account with CA (based on Offline Aadhaar e-KYC, Organisational KYC or Banking e-KYC).

6.12.3   Requirements for e-Authentication Using e-KYV Services

a) e-Sign users have unique id.
b) Application service provided should have gone through an approval process of ESP and should have agreement/undertaking with them.
c) ESP should adhere to e-KYC compliance requirements independently.
d) e-Sign user account with CA should be as per section 9.

6.13   DIGITAL SIGNATURE CERTIFICATE ISSUANCE

For a minimum of seven years (The Information Technology "Certifying Authorities"), keep a record of all pertinent information pertaining to the e-authentication of e-Sign users for the creation of key pairs and subsequent certification functions. This documentation is especially important for serving as proof for certification needs. An environment that is secure should be used to maintain such electronic records.

6.13.1   Creation of a Digital Signature

For a minimum of seven years, keep a record of all pertinent information pertaining to the e-authentication of the e-Sign user in order to gain access to the key pair. This is especially important in order to furnish proof of the creation of the digital signature. Such an electronic record ought to be kept in a safe environment.

6.13.2 Digital Signature for Indian Citizens – Class 3

Nowadays, a lot of documentation is sent electronically, and digital signatures aid in proving the sender's identity. DSC is used to validate online transactions, including Income Tax E-Filing and Company or LLP Incorporation.

6.13.3 Obtaining a Class 3 Digital Signature Is Mandatory in the Following Cases

1. E-filing income tax returns in case of every registered trust, partnership firm, companies, any other entity, or individual who is required to get tax audit under the Income Tax act.
2. Company filing with Ministry of Corporate Affairs (MCA).

MCA has mandated digital signature for the following individuals:

(i) Directors.
(ii) Auditors.
(iii) Company secretaries – whether in job or practice.
(iv) Bank official – for registration and satisfaction of charges.
(v) Other authorised signatories.

6.14 CERTIFYING AUTHORITIES

Certification agencies are appointed by the office of the Controller of Certification Agencies (CCA) under the provisions of IT Act, 2000, in India. There are a total of 14 certification agencies authorised by the CCA to issue the digital signature certificates (www.cca.gov.in). In India, digital signature certificates are issued under section 24 of the Indian IT Act, 2000, and the below certifying authorities has been given a licence to issue these certificates. One can procure class 2 or 3 certificates from any of the certifying authorities (Table 6.1).

There are a total of seven certification agencies authorised by the CCA to issue the digital signature certificates. As per the CCA 2.0 guideline 2.0 for crypto tokens, only new series of token can be used to download certificate from July 1, 2023. So old series token cannot be used to download new certificate from July 1, 2023.

6.14.1 e-Mudhra Digital Signature Online

From e-Mudhra both individual and organisations can obtain digital signatures online with ease and without using physical documents. Anyone who is interested in purchasing digital signatures can use this service. e-Mudhra is a reputable certifying

TABLE 6.1
Certifying Authorities in India

S. No	Licenced Certifying Agencies	Email Information
1	Safescrypt CA	Shankara[dot]narayanan[at]sifycorp[dot]com
2	nCode CA	ithead[at]ncode[dot]in
3	eMudhra CA	sales[at]emudhra[dot]com
4	Capricorn CA	sales[at]Certificate[dot]Digital
5	Verasys(Vsign CA	huzefa[at]thanawala[at]verasys[dot]in, Madhumita[dot]harshe[at] verasys[dot]in
6	IDSign CA	info[at]idsignca[dot]com
7	Pantasign CA	info[at]pantasign[dot]com
8	XtraTrust CA	info[at]xtratrust[dot]com,
9	ProDigiSign CA	support[at]prodigisign[dot]com, devendra[dot]singh[at]prodigisign[dot]in
10	SignX CA	info[at]signxca[dot]com
11	Care4Sign CA	info[at]care4sign[dot]com
12	RISL (RajComp) CA	kumardeepak[dot]doit[at]rajasthan[dot]gov[dot]in,
13	Protean(NSDL e-GOV)	esignhelp[at]proteantech[dot]in, nishan[at]proteantech[dot]in
14	CDAC CA	ess[at]cdac[dot]in,jahnavib[at]cdac[dot]in

authority approved by the Indian government that produces digital signature certificates for various official purposes. It is very easy to apply for a digital signature using Aadhaar OTP or PAN-based e-KYC thanks to our simplified online application process, which makes the process easier and more accessible.

To obtain DSC, applicants can easily purchase digital signatures online and complete the three-step e-KYC procedure. The entire procedure is done online and without paper.

Reliability of e-signature in Indian Law:

a. The signature creation data or authentication data are, within the context in which they are used, linked to the signatory or, as (the case may be, the authenticator and to no other person i.e. the signature must be unique to the signatory).
b. The signature creation data or the authentication data were, at the time of signing, under the control of the signatory or as the case may be, the authenticator and of no other person.
c. Any alteration to the electronic signature made after affixing such signature is detectable.

d. Any alternation to the information made after its authentication by electronic signature is detectable.
e. There should be an audit trail of steps taken during the signing process, and
f. The signer certificate must be issued by a certifying authority (CA) recognised by the Controller of Certifying Authorities appointed under the IT Act.

The second schedule provides that an "electronic signature" or electronic record can be authenticated by using electronic authentication techniques as described in the ESEATPR. The ESEATPR provides that an "electronic signature" can be authenticated by using either of the following methodologies.

a. Aadhaar e-KYC services, or
b. A third-party service by subscriber's key pair generation, storing of key pairs on hardware security module and creation of digital signature provided that the trusted third party providing such services shall be offered by any of the licenced certifying authority.

6.15 PROHIBITED DOCUMENTS

The documents or transactions that cannot be entered into by using an electronic signature are:

a. Negotiable instruments such as promissory notes or bills of exchange other than a cheque.
b. Power of attorney.
c. Trust deeds.
d. Will and any other testamentary disposition by whatever name called, and
e. Any contract for the sale or conveyance of immovable property or any interest in such property.

Further, documents to be notarised are generally required to be physically signed before the Notary public. Documents requiring registration are also to be physically signed before the concerned registrar.

6.16 DIGITAL SIGNATURE AND ITS SECURITY

Security is the primary benefit of using digital signatures. Security features of digital signatures are listed below:

Passwords, Codes, and PINs: These are employed to confirm the identity of the signer and validate their signature. The most frequently utilised techniques are email, username, and password.

Asymmetric Cryptography: This uses a public key algorithm that encrypts and authenticates using both private and public keys.

Checksum: The validity of transmitted data is assessed using this lengthy string of letters and numbers. A piece of data is hashed using a crypto-graphic algorithm to produce a checksum. To find errors or changes, the checksum value of the calculated file is compared to the checksum value of the original file. A checksum can be compared to a data fingerprint.

Cyclic Redundancy Check (CRC): This error-detecting code and verification feature, which is a subset of a checksum, is used in digital networks and storage devices to identify changes to raw data.

Certificate Authority (CA) Validation: By accepting, authenticating, issuing, and maintaining digital certificates, CAs serve as trustworthy third parties and provide digital signatures. False digital certificates can be prevented by using CAs.

Trust Service Provider (TSP): This individual or organisation provides reports on signature validation and validates digital signatures on behalf of an organisation.

6.16.1 DIGITAL SIGNATURE ATTACKS

The following attacks are common in digital signature:

Chosen-message Attack: Either the victim is tricked into digitally signing a document they didn't intend to sign, or the attacker manages to gain the victim's public key.

Known-message Attack: The victim's communications and a key that allows the attacker to forge the victim's signature on documents are both obtained by the attacker.

Key-only Attack: The victim's digital signature can be recreated by the attacker, who merely has access to the victim's public key, to sign documents or messages that the victim did not want to sign.

6.17   DIGITAL SIGNATURE IN FILING INCOME TAX IN INDIA

The main area where the digital signature is used is in income tax returns. Below is the table of income tax filed using the e-filing option. The table has been taken from the Income Tax Department from 2013 to 2023. In India, for the firms and compa-nies, the digital signature is mandatory, so it has been steadily increasing (Table 6.2 Table 6.3).

Table 6.2 depicts the e-filing of income tax from 2013 to 2023. Every year the taxpayers are increasing, which means digital signature use is increasing. After the compulsion of digital signatures for the firms the digital signature users increased simultaneously. Figures 6.1 and 6.2 explain ITR filing year on year. Individuals are separated, and others are grouped as individuals generally do not use digital signature.

TABLE 6.2

Income Tax Filing from 2013 to 2023

SR	PAN Category	AY 2013–14	AY 2014–15	AY 2015–16	AY 2016–17	AY 2017–18	AY 2018–19	AY 2019–20	AY 2020–21	AY 2021–22	AY 2022–23
1	Individual	4,95,76,555	5,38,05,146	5,79,70,144	6,55,55,912	7,04,45,510	8,04,45,511	8,55,61,788	7,79,90,888	8,25,04,957	8,90,89,795
2	HUF	9,60,004	9,99,401	10,55,205	11,19,899	11,35,677	11,87,180	12,20,604	12,17,892	12,70,730	12,89,935
3	Firm	10,35,688	10,83,515	11,56,136	12,50,519	13,12,488	14,25,375	14,83,319	15,00,260	15,59,327	16,31,592
4	Company	7,02,828	7,46,800	7,68,206	8,10,617	8,37,597	8,86,889	9,28,333	9,61,144	10,14,535	10,77,312
5	AOP(TRUST)	2,05,758	2,17,092	2,31,781	2,53,070	2,61,531	2,84,578	3,03,708	2,93,355	3,01,893	3,05,023
6	Other AOP/ BOI	1,47,353	1,66,626	1,87,754	2,14,375	2,34,845	2,67,107	3,02,888	2,94,085	3,33,547	3,55,022
7	Local Authority	5,916	7,118	7,533	8,358	9,096	10,185	11,241	10,787	11,534	12,188
8	AJP	10,211	10,556	11,098	11,702	11,506	12,106	12,078	11,649	11,508	11,968
9	Others	183	334	485	747	1,308	2,556	3,461	3,347	3,895	4,034
	Total	**5,26,44,496**	**5,70,36,588**	**6,13,88,342**	**6,92,25,199**	**7,42,49,558**	**8,45,21,487**	**8,98,27,420**	**8,22,83,407**	**8,70,11,926**	**9,37,76,869**

TABLE 6.3

Income Tax Filing Statewise from the Year 2019 to 2023

a) No of Persons who filed Income tax returns during last four years; State /year wise

State	FY 2019-20	FY 2020-21	FY 2021-22	FY 2022-23
Andaman and Nicobar Islands	37,101	41,226	44,901	47,101
Andhra Pradesh	20,80,288	19,79,366	19,84,319	21,65,161
Arunachal Pradesh	19,642	18,848	18,091	21,581
Assam	7,76,618	7,68,231	7,73,711	8,16,137
Bihar	17,19,439	18,96,122	20,11,074	21,54,266
Chandigarh	2,65,602	2,66,428	2,67,433	2,77,594
Dadar and Nagar Haveli	31,792	31,832	32,346	33,428
Daman and Diu	20,532	21,387	21,213	21,833
Delhi	34,83,436	35,34,470	35,33,774	37,06,999
Goa	2,18,697	2,17,944	2,20,219	2,30,569
Gujarat	64,73,204	69,01,630	71,26,423	74,50,672
Haryana	24,74,079	25,83,050	27,24,889	29,45,240
Himachal Pradesh	5,26,311	5,07,118	5,27,596	5,63,171
Jammu and Kashmir	4,49,249	4,25,456	4,38,770	5,22,517
Karnataka	38,18,546	39,25,684	39,80,418	42,58,035
Kerala	16,56,177	17,08,859	17,95,967	19,73,551
Lakshadweep	4,760	3,916	4,072	4,454
Madhya Pradesh	26,06,358	27,45,469	28,38,182	29,93,536
Maharashtra	1,01,34,529	1,05,05,787	1,08,22,870	1,13,91,610
Manipur	52,135	50,372	53,615	64,661
Meghalaya	33,961	32,465	34,230	40,248
Mizoram	3,808	4,885	5,866	7,371
Nagaland	20,238	20,476	20,707	25,168
Orissa	10,98,781	11,47,974	11,96,655	12,90,397
Pondicherry	97,026	95,911	95,429	1,01,440
Punjab	30,73,506	31,05,578	32,84,421	36,09,942
Rajasthan	41,35,462	43,80,416	45,55,909	48,48,031
Sikkim	14,962	13,325	11,917	13,229
Tamilnadu	41,82,347	42,07,105	43,01,299	45,90,531
Tripura	82,049	78,510	79,879	87,434
Uttar Pradesh	60,08,980	64,17,665	66,53,883	71,65,746
West Bengal	40,88,477	42,45,242	43,64,849	45,56,394
Chhattisgarh	10,41,310	10,75,894	11,01,801	11,60,389
Uttaranchal	7,69,055	7,69,961	7,93,801	8,53,992
Jharkhand	10,76,078	11,12,676	11,35,746	11,95,551
Telengana	21,58,703	22,81,927	24,54,797	26,92,185
Outside India	55,234	83,180	1,06,065	1,10,691
Others	22	3	20,246	18,191
Total	**6,47,88,494**	**6,72,06,388**	**6,94,37,383**	**7,40,09,046**

Note:

1. Unique PAN count were considered for specific FY from IT returns.

2. E-filed ITRs were considered for the above summary. In case multiple e-returns were submitted by an assessee, then the latest one in the corresponding FY has been taken into consideration.

3. Outside India – Filers who have mentioned a State Code as 99 (i.e. State outside India) in the communication address within Part A – General Information of ITR.

4. *Others – Filers who have mentioned a State Code which is inconsistent with State Codes available in Part A - General Information of ITR /PANdata.

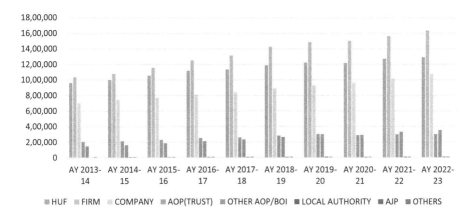

FIGURE 6.1   ITR filing (except individuals).

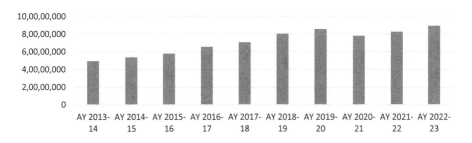

FIGURE 6.2   ITR filing for individuals from the year 2013 to 2024.

6.18   GROUNDS WHERE DIGITAL SIGNATURES ARE USED

Digital Signatures are of great importance in approving an e-document. In the following cases, digital signatures are a must:

- To digitally verify and authenticate emails.
- Security in online financial transactions.
- A signature is required in e-filing of income tax returns.
- For approving documents in MS Excel, MS Word, and PDF with a signature.

6.19   SURVEY AND FINDINGS

The following part of the study discusses the survey, information about how and why the survey was conducted, and explains the pattern of conducting interviews, analysing the process, and the results obtained from it.

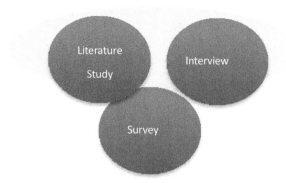

FIGURE 6.3   Details of survey. Source: Authors.

6.19.1   SURVEY

A survey was performed in India, especially in Bangalore, which is the "Silicon Valley of India," where majority of the population is aware of digital signature. The survey was conducted to support the theoretical study. The survey questions were formed by adopting "The Survey Kit: How to Ask Survey Questions" 2nd edition by Arlen Fink. Apart from this, the suggestions were taken from "Survey Monkey.com" too. The survey was conducted among non-technical, and technical, general public, but not illiterate people. Only simple questions were prepared and distributed, consisting mainly of yes/no type questions. Apart from these, multiple-choice questions were also asked.

The survey was conducted to find the knowledge about digital signatures among people from India. Around 150 people were taken randomly to answer the questions from various parts of India, especially from Bangalore. Most of the data were received by asking the participants to fill up the questionnaire and through interviews. Apart from this, the participants' data were collected through telephonic interviews, Google Meet, and Microsoft Teams using the online software named "Qualtrics." The same questions were asked in the interviews as well. In fact, some questions were skipped based on the answers given to the previous questions.

6.19.2   CHOOSING THE PARTICIPANTS

The population and the allowable error in the statistics determine how many people participate in the survey. Owing to the time constraint and the fact that the survey serves as support for the theoretical investigation, the researcher selected 150 participants Figure 6.4 gives a clear idea of how the participants were selected.

6.19.3   FINDINGS OF THE SURVEY

The questions were made simple as they were distributed to a diverse group of people such as educated and illiterate people. The survey's aim is to collect data that provide information for the following.

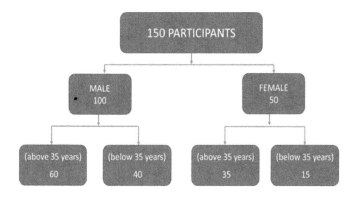

FIGURE 6.4   Choosing of participants . Note: people from below 35 years old are also above 18 years old. The people categorised under illiteracy category have not been interviewed or surveyed. Source: Authors.

Internet users versus population.
Internet users according to the survey.
Digital signature users.
Where did they get the digital signature.
How do they describe the importance of digital signature.

6.19.4   INTERNET USERS

India's population has been consistently increasing from 2020 to 2023 and has reached 1.42 billion. Simultaneously, internet users are also increasing which is a welcoming sign. Figure 6.5 shows the population versus internet users.

Figure 6.6 shows the number of internet users in different age category.

In male there are 39 people among 60, who is above 35, are internet user. But when we take the statistics of people below 35, 28 out of 40 were internet users. Among females, there are 26 people among 35 who is above 35 and are internet

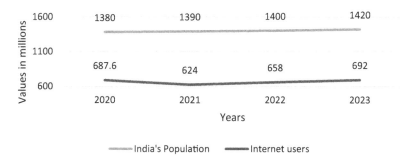

FIGURE 6.5   India's population vs internet users. Source: Authors.

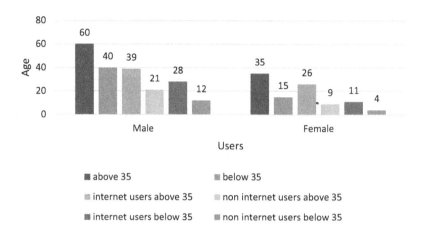

FIGURE 6.6   Internet users of different age category. Source: Authors.

users. But when we take the statistics of people below 35, 11 out of 15 were internet users. This data was collected to know how many people are not using digital signatures but are internet users. So, when the awareness of digital signatures increases, these people can be expected to become a user.

The survey also provides information about how frequently the participants use the internet. The frequency of internet usage among Swedish participants was comparatively higher than the Indian participants. Most of the participants from India use the internet almost daily in a week, mostly for checking emails and social networking.

6.19.5   DIGITAL SIGNATURE USERS

Figure 6.7 shows the number of digital signature users among the selected participants. In India, 32 people out of 150 use digital signatures. Most of the digital signature users are from the below 35 age group. Below 35 years of age, 25% of the people use digital signature, where above 35, only 20% use digital signature. This is not surprising because the internet users are more in the below 35 age group than in the above 35 age group under India.

The survey also provides information about where digital signatures are used by the participants. In India, the digital signature users mostly used it for tax filing. They also used it for sending important emails and signing documents. The people using digital signature for signing emails and documents are working in multinational companies. According to the theoretical study it was 21% out of 150 people who use DS (who is educated and above 18). The accuracy rate can be increased if the number of participants increased; unfortunately, due to the time limit we resistance towards the error now.

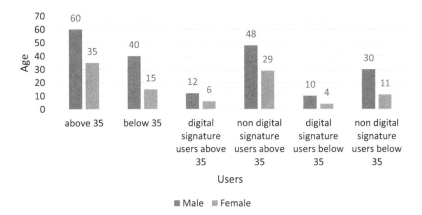

FIGURE 6.7   Digital signature users in different age category. Source: Authors.

6.19.6   KNOWLEDGE OF DIGITAL SIGNATURE

Figure 6.8 shows participants' knowledge about digital signature. From 150 participants 32 were digital signature users and 48 people have knowledge about it and the remaining were non-internet users and have no knowledge about digital signature.

The survey gives the information about the knowledge of people about digital signature. The participants are asked about where the digital signature is used? For which most of the participants answered "To authenticate in emails" and some answered for ITR purpose also. Some other said for validation of documents. Some participants are chartered accountants so they were fully aware of the uses. Participants are aware of digital signature, but do not have complete knowledge where and how digital signatures are used. Each participant was asked about the importance of digital signature in today's world, and the people with digital signature knowledge answered that evolving technology enables safety and time saving also. Even in banks the digital signatures' usage in KYC forms were also discussed. Those who don't use it answered DS as a complicated one and they don't prefer it.

6.19.7   INTERVIEWS

Interviews were done for five Fintech companies which are registered. All the registered companies know about digital signature which is mandatory. Since the companies were Fintech companies they were established recently after the digital signature was made mandatory.

This chapter has provided a clear view on the historical development of the digital signature and its laws. It has also provided detailed information about digital signature, its work, and what makes it more secure and trustable. The chapter has been done by

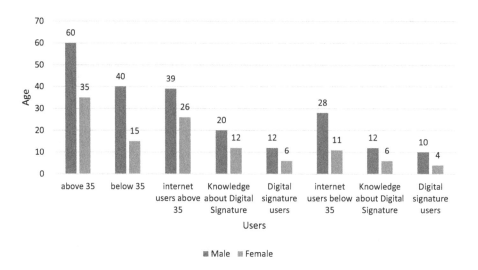

FIGURE 6.8   Knowledge about digital signature. Source: Authors.

analysing through theoretical study and practical study. The practical study was done to support the theoretical study and done by interviews and survey. It provides a clear view on the awareness of digital signature in developing and developed countries.

6.20   AWARENESS OF DIGITAL SIGNATURE

In developed countries the internet users are nearly 94% and active digital signature users. And when the population above 18 years old increases, the percentage of digital signature users also simultaneously. increases. In India the internet users are nearly 48.73%, that is, around 692 million people out of 1.42 billion population according to Digital India 2023. In the 692 million, digital signature users are around 93 million people (www.incometaxindiaefiling.gov.in). Hence 14% of the internet users in India uses digital signature. Thus, only 6.54% of the population of India uses digital signature. This proves that the awareness about digital signature is poor among the people in India and also poor among the internet users in India. The number of service available for using digital signature is also very less when compared to developed nations. Thus, the awareness of digital signature in developing country is less than the developed country. And the number of services available is less in developing countries than in developed countries.

6.21   IMPROVEMENTS TO INCREASE USAGE
OF DIGITAL SIGNATURE

In most of the countries, like USA, Sweden etc. digital signature is available for the public through banks. It can be made available by downloading from the bank website using internet banking. The banks have all the clear authenticated information about the digital signature to disseminate to the general public. But in India, there are only 14 certifying authorities (refer Table 6.1) where the public can get the digital signature (Class 3 Digital signature). Apart from this, for class 2 digital certificate pan card with video calling is needed. The usability of digital signature can be increased by making the product available in places where general public has more access, like banks and post offices. The D-mat account access is given to all the banks because of which the public investing in stock market considerably increased. Apart from this, the improvement in services of digital signature enables the people to buy more digital signature. When the usage increases the awareness also increases.

6.22   FUTURE WORK

The developing countries can follow the developed countries in implementing the digital signature in their nation. The size of the worldwide digital signature market was estimated to be around USD 5.9 billion in 2023 and is expected to increase at a compound annual growth rate (CAGR) of 40.98% to reach approximately USD 129.82 billion by 2032. The factors driving the market demand were the growing requirement for safe and legally binding electronic signatures, government laws, and the expanding necessity for electronic documents. Geographically speaking, North America leads the market due to factors like the growing use of digital technologies, the requirement for safe and effective document management, and the presence of significant players like Adobe Inc., DocuSign Inc., OneSpan Inc., and HelloSign Inc.

The industry that offers electronic signature solutions for validating and signing electronic documents is known as the "digital signature market." Cryptographic techniques are the foundation of digital signatures, ensuring the authenticity and integrity of the signed document. Due to the growing need for safe and enforceable electronic transactions across a range of sectors, including government, banking and financial services, healthcare, and education, the market for digital signatures is expanding quickly. A few of the many advantages of technology are its affordability, cost-effectiveness, and ease. Physical signatures are no longer necessary because digital signatures may be signed at any time and from any location using a variety of gadgets, including computers, tablets, and smartphones. Additionally, they provide excellent security since the digital signature's cryptographic methods guarantee that the text, paper, and document haven't been altered.

6.23   KEY MARKET CHALLENGES

Digital signatures are electronic signatures that use encryption techniques to authenticate the signer's identity and the signed document's integrity. They offer a secure and efficient way to sign and transmit electronic documents, but a lack of standardisation hinders their widespread adoption. Multiple digital signature formats exist, and different countries have different legal frameworks for recognising and regulating digital signature.

Due to the time constraint the study was conducted only in India, otherwise at least two developed countries and two developing countries would have been taken and more data would have been gathered to have a comparatively clearer idea about other countries.

REFERENCES

Acharya, M. (2024). Digital signature certificate – DSC benefits, how to get DSC, classes, download. *ClearTax*. Available at: https://cleartax.in/s/digital-signature-certificate-get-dsc

Business Today Desk. (2023). Record 6.77 crore ITRs filed for FY23 till July 31: Income tax dept. *Business Today*. Available at: https://www.businesstoday.in/personal-finance/tax/story/record-677-crore-itrs-filed-for-fy23-till-july-31-income-tax-dept-392318-2023-08-01 (updated on 1/08/2023)

Cabanellas, Guilermo. (2018). *Cyber Law in Argentina*. Wolters Kluwer: Alphen aan den Rijn, Netherlands.

Diffie, W., and Hellman, M.E. (1976). New direction in cryptography. *IEEE Transaction on Information Theory*, 22(6): 644–654.

Fink, A. (2002). *The Survey Kit: How to Ask Survey Questions*. 2nd ed. Sage Publications.

Goldwasser, S., Micali, S. and Rivest, R. L. (1988). A digital signature scheme secure against adaptive chosen-message attacks. *SIAM Journal on Computing*, 17(2).

Government of India. (2023). IRT e-filing. Available at: https://incometaxindiaefiling.gov.in/. (Last checked 2023-9-01).

Kemp, S. (2023). Digital 2023: India. *DataReportal*. Available at: https://datareportal.com/reports/digital-2023-india#:~:text=Internet%20use%20in%20India%20in,unchanged%20between%202022%20and%202023.

Kessler, G. C. (1998). An overview of cryptography. *Handbook on Local Area Networks*. Auerbach Publications.

Latimer, P. (2017). Signatures, squiggles and electronic signatures. *SSRN*. Available at: https://ssrn.com/abstract=1601169

Mason, S. (2006). Electronic signature in practice. *Journal of High Technology Law*, 148–164.

MCA. (n.d.). FAQs on digital signature certificate (DSC). *Ministry of Corporate Affairs*. Available at: https://www.mca.gov.in/MinistryV2/digitalsignaturecertificate.html (last checked 23rd November 2023)

Panko, R. (2004). *Corporate Computer and Network Security*. Prentice Hall: Hoboken, NJ.

Precedence Research. (2024). Digital signature market size, share, and trends 2024 to 2033. *Precedence Research*. Available at: https://www.precedenceresearch.com/digital-signature-market

Rivest, R.L., Shamir, A. and Adleman, L. (1977). A method for obtaining digital signatures and public-key cryptosystems. *Communications of the ACM*, 21(2), 120–126.

signNow. (n.d.). Legality and enforceability of electronic signatures in India. *signNow*. Available at: https://www.signnow.com/legality/india?gad_source=1&gclid=CjwKCAiApaarBhB7EiwAYiMwqmHgZFzaJ14N2WGA3zwWcmnU5QKUMP2WnajKkzLXkbF5WkiUd19NxRoC7CsQAvD_BwE

7 AI-Infused Blockchain Technology for Thrust Applications

C. Venkataramanan, S. Dhanasekar,
V. Govindaraj, K. Martin Sagayam, and P. Geetha

7.1 INTRODUCTION

Blockchain is a database constructed in a distributive manner to hold the data immutably (Verma et al., 2022). The interconnected blocks are securely connected to each other and used to track and build trust. If any transaction needs to be updated in a block, it can't be altered without altering the succeeding blocks. By definition, it maintains a list of arranged data records (i.e. blocks) contains a information about its previous section details (Zerka et al., 2020). Digitalization will encompass the majority of engineering fields, with a bigger impact on wide-area communication networks involving fast data transmission (Dhanasekar et al., 2022; Dhanasekar et al., 2021; Neebha et al., 2023). The chain-like construction is a decentralized one, and the data can be recorded across the region of interest, and the same cannot be altered without impacting the subsequent blocks (Raja et al., 2022; Firouzi et al., 2022). The blockchain records the information digitally. The main difference between the blockchain and a conventional database is that in blockchain, the data is stored in a structured manner. It gathers the information in groups (i.e blocks). Each block has a storage capacity, and it gets connected with its previous one once filled, forming the chain (Govindaraj et al., 2023; Bruntha et al., 2022). Every piece of information in the freshly filled blocks is connected to its previous one, continuing the chain, which is called the blockchain. Whenever a block is associated with the chain, an exact timestamp is given (Firouzi et al., 2023). Without any assistance from a third party, the blockchain technology ensures the integrity and security of the data.

It is an advanced database technology that allows the information sharing in a transparent manner within a network. Data originating from many sources are collected and shared using the blockchain cloud environment (Bansal et al., 2022). Each block in the chain contains a unique cryptographic hash ID, which avoids the duplication of data. The main feature of this technology is that data can't be transacted without the knowledge of the associated parties (Wazid et al., 2023).

DOI: 10.1201/9781003453109-7

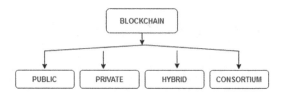

FIGURE 7.1   Types of blockchain. Source: Authors.

7.1.1   Types of Blockchain Technology

In general, four different types of blockchain technologies exist: public, private, hybrid, and consortium blockchain (Figure 7.1).

• **Public Blockchain**

As the name implies, anyone can join in this non-restrictive distributed ledger environment and perform transactions without any permissions. Each user in this type has the authority to conduct mining operations. The necessary verifications are completed to perform the complex operations before adding them into the register. Since this type is available to the public, anyone can check and verify the transactions.

• **Private Blockchain**

In general, this private blockchain exists in an organization or firm (i.e. limited network). Unlike public blockchain, this type operates in private circumstances and has the controlled environment, since no one can join without permission. It is also called a permission-oriented blockchain and business blockchain.

• **Hybrid Blockchain**

It is the combination of public and private blockchain types. The best features of both are combined together to provide services. In this type, initially the records are not available publicly and the access can be granted upon validation.

• **Consortium Blockchain**

Like a hybrid type it has the features of both public and private types. Here, the control is taken by the predetermined peers for consensus methods. It slightly differs from the hybrid in that it involves different organizational members. A special purpose node is available to validate the transactions from the members.

7.1.2   Necessity of Different Types of Blockchain

The situations and environments are different from each other; to carry out transactions securely in such areas requires different types of blockchain methods. Apart from this, anyone can validate others from different geographical regions and exchange information. To access the network with different authentication methods are also the need for various types of blockchain methods. According to user requirement the types of blockchain networks can be set up along with necessary services.

7.1.3 Applications of Technology

- **Money Transfer**: Payment transactions over the blockchain can be made easier and completed within a short duration securely.
- **Supply Chain Environment**: Using this technology, the lagging points in the business can be easily identified from manufacturers to suppliers.
- **Digital Authentication**: People can identify digitally and provide authentication to others who need to utilize.
- **Distribution of Data and Security**: It allows people to share their data among themselves securely.
- **Intellectual Property Rights**: They are used to protect the findings and ensure transparent utilization and royalty distributions.
- **Internet of Things**: The blockchain network monitors and identifies the devices to be newly added and determines whether they are trustworthy or not.
- **Agriculture**: The technologies IoT (Internet of Things) and blockchain are combined together and used to create a secure digital forming system which ensures food safety, quality, and production.
- **Healthcare**: Blockchain plays a vital role in healthcare, such as maintaining the patient records and clinical trials.
- **Technologies Used in Blockchain**: There are three main technologies playing a vital role when designing the blockchain: cryptographic hash keys, a network with a shared ledger, and computing.

7.1.4 Working Principle of a Blockchain

The entire life cycle of a blockchain process can be explained in the following six steps:

1. Initialization of a transaction.
2. Verification process.
3. New block formation.
4. Consensus mechanism.
5. Addition of new block.
6. Transaction completion (Figure 7.2).

1. **Initialization of a Transaction**

It is the process of the beginning of a transaction (i.e. a new record enters into the blockchain). The information records need to be shared are encrypted with the cryptographic hash keys, either public or private keys.

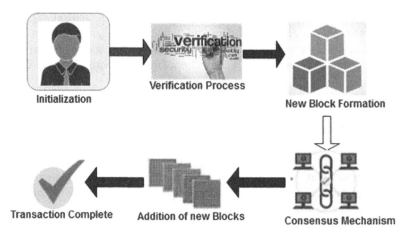

FIGURE 7.2   Life cycle of the blockchain process. Source: Authors.

2. **Verification Process**

The entered record is then shared into the blockchain distributed network across the region of interest. The respective peers in the network verify its resources to effectively carry out the given transaction.

3. **New Block Formation**

At the given time, a greater number of transactions are verified and identified as legitimate. Then, the legitimate transactions are combined together and form the block.

4. **Consensus Mechanism**

The block formed by the legitimate sources is tried to be added to the blockchain network. If all the peers simultaneously try to do the process, it causes the working of the blockchain. To address this issue, the peers use the consensus mechanism. The consensus mechanism allows the peers to add the block only once to the network securely and ensures the process is not done by other peers. The source that is involved in the consensus process is called as miners. The consensus algorithm generates the hash code for the intended block and adds the same to the blockchain.

5. **Addition of New Blocks**

Once a block gets its hash key and is authenticated, it can be added to the blockchain. Every block of the blockchain has information about the previous one and details about how the intended block is connected with others in the blockchain. The newly formed block can be added to the network openly and end of it.

6. **Transaction Complete**

Once the block gets connected to the blockchain network, the record transaction process is completed, and the entire details about the transaction are permanently shared in the blockchain network. Whoever is interested can retrieve the records shared in the network and confirm them.

7.1.5 INTEGRATION OF AI AND BLOCKCHAIN

The combination of artificial intelligence (AI) and blockchain plays a vital role in many industries and provides necessary improvements where they have been implemented. The major examples are healthcare, supply chain, media, and financial services. The combination protects the industry environment against cyberattacks too. AI effectively handles datasets, even sizes as large and make the output patterns relate to the behavior of input data. The integration of blockchain with AI is used to eliminate the duplications and bugs. The enhanced classifiers and patterns of AI can be easily verified by the blockchain network for authentication. This kind of process is easily adopted and used in the customer-faced business and transactions. The integration of AI with blockchain is used to automate the marketing in blockchain infrastructure. The influence of AI in blockchain technology ensures a reliable decision-making system and provides secure transactions. It offers many benefits such as generalized verification systems, enriched data models, smart retail systems, efficient predictive analysis, transparent monitoring, financial services, and so on.

AI-enabled blockchain technology ensures security in various applications. One example is that the decision-making capabilities of AI algorithms are enriching the performance of blockchain in terms of security in financial transactions and can block or investigate falsified ones. During the consensus process, AI optimize the calculations and minimizes the burden on miners, which greatly reduces the delay for respective transactions. The processing cost as well as the energy spent by miner peer considerably reduced because of the assistance given by AI. When data records increase simultaneously, the AI algorithms are applied on to the blockchain data to optimize it. AI uses decentralized algorithms and new sharing techniques, which make the blockchain network much more efficient.

The AI-implemented blockchain increases the trust between the peers in the network, which will enhance the integrity of the data records shared with each other. The AI engine makes the decision and shares the data even when the records are larger in size. AI manages the blockchain networks by training the records to make decisions accurately. The AI algorithms are used to enhance the quality of data records from the database compared to a human expert with considerable practice. Data security is one of the major challenges when it is available in public. Homomorphic algorithms are used in general to encrypt the records directly. This way, each user can get the advantage of privacy in the blockchain network. The blockchain network is one of the better ways to store the personal data since it is securely processed by AI, which makes the blockchain network more efficient and convenient.

7.1.6 AI-INFUSED BLOCKCHAIN

AI-infused blockchain enables the blockchain network to be accessed securely from inside and outside of the organization. The management process of the blockchain network, such as records consumption, data management and arrangement, sharing procedures, transparency, and trust are enhanced when AI integrated with the blockchain network model.

- **Power Management**

When the blocks are attempted to crack more power will be consumed since the entity has applied algorithms like brute force to ensure integrity. The number of blocks to be processed is also in higher order. If the processing methodology is composed of AI algorithms (i.e. machine learning algorithms), the data records can be easily mined (i.e. training process of the records) and provide solutions that can minimize the processing power for the entire operation.

- **Enabling API**

The blockchain approach creates the database in an organized manner. The networks are highly transparent and made easily accessible to everyone around the world. By enabling the API which in turn interacts with the AI module (i.e. containing various learning algorithms) to process the records efficiently.

- **Security**

The financial and medical records are highly sensitive and need to be kept very safely. They need to be stored in public and shown to the parties when verified. Blockchain technology stores the information very securely in the form of encrypted records. By enabling the AI, it continuously updates itself and feeds the knowledge sets, giving more advantages.

- **Decision-Making**

The records, when entering into the blockchain network to the conclusion step, are to be highly secured and not easily tampered with. It can make trust with the blockchain technology, and the AI-based machine learning algorithms make the process easier when the records are larger in size by optimizing them. This makes AI-based decision-making in the blockchain network more interesting. The motivation of this literature is to provide insight in the necessities of AI in blockchain technology. It provides a clear understanding when incorporating the AI into blockchain. Furthermore, it details the major technological advancements in the area as well as open research challenges in the field of blockchain technology with AI.

7.2 THRUST AREAS OF AI-ENABLED BLOCKCHAIN

The major thrust areas to be concentrated are healthcare, agriculture, supply chain, and IoT. The combination of AI and blockchain transform the listed applications into the next era. The detailed approaches to the above listed applications are analyzed below in detail (Figure 7.3).

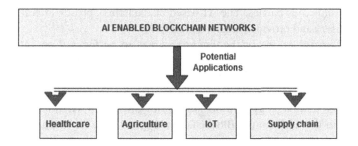

FIGURE 7.3 Thrust applications of AI-enabled blockchain. Source: Authors.

7.2.1 BLOCKCHAIN AND AI IN HEALTHCARE

The algorithms designed in AI are used to enhance decision-making according to the data fed into them. In healthcare industry, AI algorithms are used to gather data, analyze it, and make decisions according to the algorithms designed. The analysis and output presentation provide optimized results through significant patterns with reduced errors. Past history and symptoms are fed as input to the AI-based algorithms to predict the disease as output. The output can be given in the form of alerts, warnings, and care tips for the data given as input. In medical imaging systems, AI algorithms can detect even the minute changes that can't be recognized by human beings directly. One of the widespread applications of AI is telemedicine. The programs developed in the system can monitor patients and convey the necessary information to the concerned person to take immediate action and improve the success rate.

One of the emerging techniques in the healthcare industry is blockchain technology, which can store data in a shared ledger securely and in an organized manner. The utilization of fake drugs is noticeably reduced with the help of a blockchain network. The combination of blockchain and AI can transform the healthcare industry to the next level.

The proof of work and the proof of stake are the consensus algorithms used widely in the blockchain network and will not consider the computational complexity and other metrics (Kumar et al., 2020). The fairness of the blockchain network is enhanced by implementing AI into the consensus algorithms. A healthcare setup has been established to produce the dataset and applied input to the network. The role of the miner peer is to substantiate the data among others. The disadvantages of the miner nodes are noticeably discarded by using the AI-based blockchain consensus model, which can further improve the performance of the blockchain network.

One of the major issues of the health care industry is fraudulent claims of health insurance (Kapadiya et al., 2022). At present, most of the people are engaged with health insurance to face sudden expenditures (i.e. health emergency) if any. Nowadays, the healthcare insurance industries are facing many challenges like security, fraud, and privacy. Organizations need to come forward to develop systems to act against malicious parties. The AI-based blockchain approach effectively diminishes fake users and provides secure transaction.

Digital farming is one of the emerging areas to be focused on currently. The Internet of Things (IoT) is playing a major role in smart agriculture (Fan et al., 2021). The area of interest is continuously monitored and collected for further processing. However, accuracy and reliability are the two main parameters deciding the decision. Conventional data aggregation and privacy protection will not solve the issues completely in digital farming. Blockchain, along with the AI module, provides the best effort solution and decentralized architecture to enhance fault-tolerant capability in digital agriculture.

The AI-enabled blockchain network was implemented in agricultural supply chain management. AI-based algorithms enhance the quality of service and efficiently use the resources. Unmanned aerial vehicles (UAVs) combined with the AI and blockchain are used in agriculture for tracing, tracking, and contracts (Zawish et al., 2022). Instead of a single compressed convolutional neural network, the fully convolutional neural network (FCN) is used to handle more than one task-specific model with less complexity and improved accuracy.

AIoT enabled novel blockchain architecture is developed and used to monitor the area of interest (Alrubei et al., 2022). The collected data are processed, analyzed, and validated and then shared into the blockchain network. During the process, the AI engine effectively handles the data and provides meaningful outcomes.

According to the expert's prediction, in the near future millions of devices can connect with the internet, out of which one-third are mobiles, desktops, and tablets (Banafa 2022). The remaining devices are sensor units, smart home appliances, and other IoT-enabled terminals. The role of IoT is to connect people, data, and processes. To provide trustworthiness among all the peers in the network, it is necessary to combine AI with blockchain and IoT. By enabling this, it can be ensured that the necessary privacy is provided among every node in the blockchain network (Gupta et al., 2021).

7.2.2 OPEN RESEARCH CHALLENGES

The open research challenges when combining artificial intelligence with blockchain networks are elaborately detailed below.

- **Privacy Protection**

The records fed into the blockchain network are publicly available in the shared ledgers and the same are available to all the readers. In parallel, the devices concerned are continuously collecting private and sensitive data and putting them onto the open shared ledgers. This is the point of concern we look into. Data privacy should be ensured by using encryption functions to allow access. The increased number of records can be effectively handled when processed with the AI to make the necessary decisions.

- **Scalability and Side Chains**

Scalability is one of the important parameters when considering any kind of network. For example, to date 4 and 12 transactions are performed by the Bitcoin and Ethereum blockchain networks, respectively. Facebook handles millions of transactions every second. To enhance the process of blockchain, side chains are used. Transactions made between the two parties are sorted out quickly outside the main chain and updated once a day in the main chain. Even though more algorithms are available to improve the performance, it still it needs more algorithms to address scalability issues.

- **Security**

The new trusted execution environments are needed to avoid the tampering issue in the miner nodes of blockchain networks that contain the hyper ledgers. The system using the public ledger type suffers (cyberattacks) more, about 51% compared to the private type. The private type uses predefined consensus protocols among the peers. In this view, the security is the major issue in the blockchain network that needs to be focused more.

- **Smart Contracts and External Functions**

The poor programming practices used to code the smart contract between the application side and the blockchain network is one of the series issues and is vulnerable. Hence, it is necessary to protect the code as well as the records on the blockchain network. New effective testing tools also need to be developed to test the security condition of the smart contract. AI-based machine learning algorithms are used to approximate the outcomes rather than random ones. Hence, a clear step needs to be taken against the issues listed above.

- **Trusted Database**

The peers of the blockchain network simultaneously work with the external functions by using the smart contracts (i.e. database). Sometimes the data are pulled into the blockchain network and pushed to the third party. Hence, to work with the third-party external functions, trusted functions to be deployed avoid insecurity.

- **Application level – AI-Specific Consensus Algorithms**

The consensus protocols in the blockchain network act as a middle layer to validate the processes. An urgent focus is needed to concentrate on developing the consensus algorithms at the application level with enhanced learning approaches to improve performance and optimization quality.

- **Localized Computing**

The processing delay slightly increased when working with the cloud environment. Fog computing is a technique used to compute the process locally (i.e. nearer to the source). In the context of a blockchain network, the customers or IoT devices continuously generate data. If the generated records are processed locally with AI-based machine learning approaches, by using the fog peers, the delay can be minimized considerably. Hence, a necessary action needs to be taken on the development of fog nodes in the blockchain network influenced by AI.

- **Standards and Interoperability Development**

Blockchain networks have yet to be developed further. The standards organizing committee needs to ensure the blockchain interoperability with other standards. Global-level governance needs to be developed to ensure the operation of blockchain, as it completed growth with financial markets. Hence, more research should be concentrated on developing the blockchain models, services, and architectures.

- **Quantum Computing**

Quantum computing is one of the emerging research projects which can break the public key encryption standards and determine the private keys in the near future. At present, the blockchain network works with public key encryption. Hence, solid research needs to be developed on quantum-safe blockchain to withstand against such quantum computing breaches. The mitigation capability should be ensured with quantum-flexible blockchain network platforms.

- **Monitoring Agents**

One of the major challenges is to manage the blockchain network among the different stakeholders. It is very difficult to assign roles for managing the functions such as deployment, administration, smart contracts, troubleshooting, monitoring sidechains, interoperation, and so on. Hence, more research is necessary to focus on the development of governing models.

7.3 FOG COMPUTING IN BLOCKCHAIN WITH ARTIFICIAL INTELLIGENCE

Fog computing is a technique that extends cloud computing to the edge of the network and makes the network as a distributed computing infrastructure. It constructs the data storage and processing nearer to the data source, which minimizes the delay and utilization of bandwidth, and facilitates the real-time data processing. The integration of fog computing with artificial intelligence (AI)-infused blockchain can enhance the security, scalability, and overall efficiency of various applications.

How the combination of fog computing with blockchain and AI is enhancing the performance factors is explained below.

The effective utilization of bandwidth (BW) and optimum delay can be ensured by allowing the AI process at the edge of the blockchain network. Fog computing allows the processing to be done at nearer to the data source instead of sending it to a centralized cloud environment. Hence, this technique can be used widely where time-sensitive data are handled (i.e., time-bounded Internet of Things (IoT), vehicular area networks). Furthermore, the below example clearly explains how the fog computing with AI-infused blockchain minimizes delay and BW utilization in an intelligent traffic management system.

7.3.1 INTELLIGENT TRAFFIC MANAGEMENT SYSTEM

Different sensors and cameras are deployed at the roadsides to monitor the traffic conditions. The sensors in the respective region of interest collect real-time data such as the movement of vehicles, frequency of traffic, and conditions of the road. This data is used to optimize the flow of traffic, reduce congestion, and ensure road safety. In the conventional system, the collected information is stored directly on the cloud server for further processing. The cloud servers are responsible for running the AI algorithms to evaluate the data, predict patterns, and deliver the optimized instructions. The major challenges of this approach are delay and overutilization of bandwidth. The delay is incurred due to the process of transporting the data from the region of interest to the centralized server, leading to slowing down of the responses and congestion. The amount of bandwidth required to send the raw data from the field to the centralized server is significant, which, in turn increases the cost of conveying data.

The adoption of fog computing with the AI-based blockchain reduces the amount of data conveyed to the centralized server and places the fog nodes nearer to the field of interest. The fog nodes act as local computing regions to reduce the burden on centralized cloud services.

Merits of Fog Computing with the AI-based blockchain:

- Fog nodes do the data analysis locally.
- It provides real-time information about the area of interest.
- Faster decisions can be made.
- The amount of data to be sent to the server is reduced.
- Optimized BW reduces the network load.

The integration of blockchain made the network more secure and trustworthy; further, it provides immutable and tamper-proof records to ensure effective traffic management decisions. This architecture enables the fog nodes to use AI algorithms for analysis and forecasting. This arrangement optimally handles real-time traffic conditions in a smoother way and ensures congestion-free communication. In summary, the AI-infused blockchain with fog computing ensures reduced latency, effective utilization of bandwidth, security, and efficiency in traffic management.

7.3.1.1 Pervasiveness of Privacy and Security

In fog computing, the integrity and security of data are improved locally instead of sending it to a cloud server. This is one of the significant services of blockchain, which handles the personalized information. The stakeholders can rely on data privacy and integrity in blockchain technology associated with AI. The example shown below is brief about how an intelligent home automation system is enhanced with the adoption of fog computing with AI and blockchain.

7.3.2 Intelligent Home Automation System

The smart home automation system consists of various devices such as cameras, locks, and thermostats, providing seamless connectivity for a smart living environment. The devices, with respect to the area of interest, gather the data and convey it to centralized coordinator to manage smart activities. In the conventional system, the collected information is directly sent to the cloud server for further processing. The AI algorithms available in the cloud server analyze the received data to make wiser decisions regarding control and automation. The major challenges associated with smart home systems are privacy preservation, security, and so on. The smart home environment always handles sensitive information, such as video from a surveillance camera. When this information is shared with the cloud server, it leads the privacy issues that may be undergone with the unauthorized access. Another issue is tampering of the cloud server by hackers. A security breach can compromise the data as well as lead to unauthorized access to the smart home automation systems. The introduction of fog computing with intelligent home automation limits security and privacy issues.

In general, the fog nodes are deployed at the edge of the network to process the data locally instead of sending it to the centralized cloud. The data received from the sensors is processed at the network edge itself and not sent outside the network. The main advantage of these fog nodes in a smart home is that they manage different tasks separately according to the data received and processed. Along with data collection and processing, they are also involved with the actuator control mechanisms. They send the data required for a particular task only to the cloud server.

A fog node improves privacy protection by minimizing the amount of data transferred to cloud which results in increased privacy. The video information available from the security cameras remains in the local network and can be accessed by only authorized users. The fog nodes are integrated with additional features like encryption and access control, which protects the information within the network itself.

To enhance the security, the smart home network can be integrated with a blockchain-based access control mechanism, and the smart contract ensures permission across various smart systems. The authorized users can access the environment at the fog node level with limited privileges.

This approach ensures the cloud server can receive only the aggregated information from each device and not all the data, which reduces the chances of data breaches as well as unauthorized access. From the given example, it is clear that the integration of fog nodes with AI and blockchain enhances security, privacy, and efficient utilization of data records in a convenient way.

7.3.3 OFFLINE AI COMPETENCE WITH FOG COMPUTING IN MOBILE HEALTH MONITORING SYSTEM

The fog computing permits the AI to run nearby the end network devices with or without internet connectivity. This particular case is more beneficial where internet access is not required continuously or in areas where the network outages persist.

The development of handheld health monitoring devices such as smartwatches and fitness bands is used to monitor the human health in an optimal manner. It collects the vital signs of human health continuously and stores them in a central cloud system. The data stored in the cloud is analyzed, processed, and given the alerts. In the conventional process, the wearables continuously monitor the signs and send them directly to the cloud server for analysis. The AI algorithms available in the cloud environment process the data and generate recommendations. The major challenge of the conventional system is that the entire architecture is internet-dependent. It requires continuous internet support to provide optimum recommendations, and false alarm may be generated in areas of poor network.

The fog computing can be integrated into mobile health monitoring to facilitate AI competence without the internet. The fog nodes are deployed in various places to process data that belong to health monitoring locally without the need of an internet connection. The fog nodes with AI capabilities can run AI algorithms locally, analyze the health data, detect abnormalities, and provide the insights to users without sending all the information to the cloud. The fog nodes can generate alerts locally to the end users about their health conditions without any internet support. The mist nodes collect the information from the sources until an internet connection is available; when the connection is restored, the collected data is synchronized with the cloud environment, ensuring that data may not be lost at any point.

The integration of blockchain in mobile health monitoring leads the safe and secure data processing using fog nodes. Once the health-related data is entered onto the blockchain, it becomes tamper-proof and immutable, which improves the data integrity.

This arrangement permits users to receive health-related recommendations even when they are not in the area of internet coverage. It provides timely alerts to individuals and healthcare professionals with greater reliability and accessibility.

7.3.4 SCALABILITY IN SUPPLY CHAIN MANAGEMENT

The example below details how blockchain, fog computing, and scalability are combined to accomplish the goal of supply chain management.

In a global-level supply chain, the products may go through various stages involving different stakeholders, producers, suppliers, retailers, and distributors. Each part of the system generates data about the operations. In the conventional supply chain system, the data is stored in the central server and accessed by various stakeholders. Information sharing, as well as ensuring data security is maximizing the processing time. Scalability issues have increased as the volume of data in the supply chain grows. A major challenge in this category is maintaining unified data (i.e. preventing fragmentation) across the entire network. This leads to network inefficiency and transparency issues. The successive delay between transactions also increases when validating and synchronizing information.

The blockchain uses a shared and immutable ledger that is accessed by all the authorized parties of the network. Each entry is recorded as a chain of information. It ensures data integrity and transparency. The implementation of edge nodes in the blockchain is used to process the information locally and complete operations such as tracking, validation, and inventory tasks. The role of edge nodes in the blockchain is to verify new transactions before adding them to the blockchain network, which guarantees the data integrity in the network. The main advantages of introducing scalability are to provide better transparency and minimized delay. The integration of supply chain management combined with fog computing results in a scalable and efficient system.

7.3.5 Fog Computing for Real-Time Decision-Making in Smart Grid

The example below explains how the integration of blockchain and edge computing are used in real-time decision-making in a smart grid.

The smart energy grid consists of various IoT devices such as sensors, monitoring devices, and processing equipment to observe the entire area of interest in terms of energy consumption, generation, and distribution of energy. The major objective of the smart grid is to optimize the utilization of energy and ensure fluctuation handling mechanisms in distribution and requirements. In conventional systems without fog computing, the data from the monitoring devices are directly stored in the centralized cloud. Data analysis can be done at the central level for further processing. The main challenges of these systems are delay and single point failure. The inclusion of fog nodes in the blockchain-based smart grid facilitates the process of data verification and analysis locally, which reduce the burden on network and ensuring data integrity, reduced delay, and offline process too. The main advantages of fog nodes in the blockchain network are reduced delay, network resilience, and enhanced energy efficiency. They make the system dynamic, stable, and an energy-efficient one.

7.3.6 Fog Computing for Cooperative Network of Autonomous Vehicles

The example below narrates how blockchain, fog computing, and collaborative AI models can be applied for an efficient autonomous vehicle network.

When autonomous vehicles start to operate in a smart city, the majority of the vehicles need to adopt with various sensors and AI models for safe and efficient driving. The vehicles are expected to share information among each other to improve overall traffic and safety on the roads. In the conventional system, decisions are made independently based on local observations which limits smart city expectations and their benefits. The major challenges in this system are tight collaboration between the vehicles and data privacy which affects the overall performance of the network. Collaborated AI using fog computing with blockchain ensures enhanced traffic management, safety, data integrity, trust between the peers and a distributed learning environment. The integrated system provides shared communication and processing results, which enhance transportation in the smart city environment.

7.4 CONCLUSION

This chapter has made a detailed analysis of the blockchain network in combination with artificial intelligence and IoT. The features, types, technologies, and processes of the blockchain network are discussed with suitable graphical illustrations. The necessity of enabling AI into the blockchain is discussed with necessary examples. The performance enhancement of the blockchain when integrating the AI is discussed with supporting literature. Further, the potential thrust areas (i.e. Applications) of AI-infused blockchain models are clearly summarized with literature support. Finally, the open research issues and challenges are discussed, and the future directions are given. In the future, we may not expect technologies without AI with blockchain to ensure privacy and security.

REFERENCES

Alrubei, S. M., Ball, E., & Rigelsford, J. M. (2022). A Secure Blockchain Platform for Supporting AI-Enabled IoT Applications at the Edge Layer. *IEEE Access*, 10, 18583–18595. https://doi.org/10.1109/access.2022.3151370

Bansal, G., Rajgopal, K., Chamola, V., Xiong, Z., & Niyato, D. (2022). Healthcare in Metaverse: A Survey on Current Metaverse Applications in Healthcare. *IEEE Access*, 10, 119914–119946. https://doi.org/10.1109/access.2022.3219845

Banafa, A. (2022). Using Blockchain to Secure IoT. *Secure and Smart Internet of Things (IoT)*, 77–84. https://doi.org/10.1201/9781003339373-14

Bruntha, P. M., Dhanasekar, S., Hepsiba, D., Sagayam, K. M., Neebha, T. M., Pandey, D., & Pandey, B. K. (2022). Application of switching median filter with L2 norm-based auto-tuning function for removing random valued impulse noise. *Aerospace Systems*, 6(1), 53–59. https://doi.org/10.1007/s42401-022-00160-y

Dhanasekar, S., Bruntha, P. M., Neebha, T. M., Arunkumar, N., Senathipathi, N., & Priya, C. (2021). An Area Effective OFDM Transceiver System with Multi-Radix FFT/ IFFT Algorithm for Wireless Applications. 2021 7th International Conference on Advanced Computing and Communication Systems (ICACCS). https://doi.org/10.1109 /icaccs51430.2021.9441694

Dhanasekar, S., Stella, T. J., Thenmozhi, A., Bharathi, N. D., Thiyagarajan, K., Singh, P., Reddy, Y. S., Srinivas, G., & Jayakumar, M. (2022). Study of Polymer Matrix Composites for Electronics Applications. *Journal of Nanomaterials*, 2022, 1–7. https:// doi.org/10.1155/2022/8605099

Fan, H., Yang, H., & Duan, S. (2021). A blockchain-based smart agriculture privacy protection data aggregation scheme. 2021 2nd International Conference on Artificial Intelligence and Computer Engineering (ICAICE). https://doi.org/10.1109/icaice54393.2021.00018

Firouzi, F., Farahani, B., Barzegari, M., & Daneshmand, M. (2022). AI-Driven Data Monetization: The Other Face of Data in IoT-Based Smart and Connected Health. *IEEE Internet of Things Journal*, 9(8), 5581–5599. https://doi.org/10.1109/jiot.2020.3027971

Firouzi, F., Jiang, S., Chakrabarty, K., Farahani, B., Daneshmand, M., Song, J., & Mankodiya, K. (2023). Fusion of IoT, AI, Edge–Fog–Cloud, and Blockchain: Challenges, Solutions, and a Case Study in Healthcare and Medicine. *IEEE Internet of Things Journal*, 10(5), 3686–3705. https://doi.org/10.1109/jiot.2022.3191881

Govindaraj, V., Dhanasekar, S., Martinsagayam, K., Pandey, D., Pandey, B. K., & Nassa, V. K. (2023). Low-Power Test Pattern Generator Using Modified LFSR. *Aerospace Systems*. https://doi.org/10.1007/s42401-022-00191-5

Gupta, R., Shukla, A., & Tanwar, S. (2021). BATS: A Blockchain and AI-Empowered Drone-Assisted Telesurgery System Towards 6G. *IEEE Transactions on Network Science and Engineering*, 8(4), 2958–2967. https://doi.org/10.1109/tnse.2020.3043262

Kapadiya, K., Patel, U., Gupta, R., Alshehri, M. D., Tanwar, S., Sharma, G., & Bokoro, P. N. (2022). Blockchain and AI-Empowered Healthcare Insurance Fraud Detection: an Analysis, Architecture, and Future Prospects. *IEEE Access*, 10, 79606–79627. https://doi.org/10.1109/access.2022.3194569

Kumar, N., Parangjothi, C., Guru, S., & Kiran, M. (2020). Peer Consonance in Blockchain based Healthcare Application using AI-based Consensus Mechanism. 2020 11th International Conference on Computing, Communication and Networking Technologies (ICCCNT). https://doi.org/10.1109/icccnt49239.2020.9225550

Neebha, T. M., Andrushia, A. D., Malin Bruntha, P., Varshney, A., Manjith, R., & Dhanasekar, S. (2023). On the Design of Miniaturized C-Shaped Antenna Based on Artificial Transmission Line Loading Technique. *Journal of Electromagnetic Waves and Applications*, 37(6), 814–826. https://doi.org/10.1080/09205071.2023.2211282

Raja, L., & Periasamy, P. S. (2022). A Trusted Distributed Routing Scheme for Wireless Sensor Networks Using Block Chain and Jelly Fish Search Optimizer Based Deep Generative Adversarial Neural Network (Deep-GANN) Technique. *Wireless Personal Communications*, 126(2), 1101–1128. https://doi.org/10.1007/s11277-022-09784-x

Verma, A., Bhattacharya, P., Madhani, N., Trivedi, C., Bhushan, B., Tanwar, S., Sharma, G., Bokoro, P. N., & Sharma, R. (2022). Blockchain for Industry 5.0: Vision, Opportunities, Key Enablers, and Future Directions. *IEEE Access*, 10, 69160–69199. https://doi.org/10.1109/access.2022.3186892

Wazid, M., Das, A. K., Shetty, S., Rodrigues, J. J. P. C., & Guizani, M. (2023). AISCM-FH: AI-Enabled Secure Communication Mechanism in Fog Computing-Based Healthcare. *IEEE Transactions on Information Forensics and Security*, 18, 319–334. https://doi.org/10.1109/tifs.2022.3220959

Zawish, M., Ashraf, N., Ansari, R. I., Davy, S., Qureshi, H. K., Aslam, N., & Hassan, S. A. (2022). Toward On-Device AI and Blockchain for 6G-Enabled Agricultural Supply Chain Management. *IEEE Internet of Things Magazine*, 5(2), 160–166. https://doi.org/10.1109/iotm.006.21000112

Zerka, F., Urovi, V., Vaidyanathan, A., Barakat, S., Leijenaar, R. T. H., Walsh, S., Gabrani-Juma, H., Miraglio, B., Woodruff, H. C., Dumontier, M., & Lambin, P. (2020). Blockchain for Privacy Preserving and Trustworthy Distributed Machine Learning in Multicentric Medical Imaging (C-DistriM). *IEEE Access*, 8, 183939–183951. https://doi.org/10.1109/access.2020.3029445

8 The Nature of Trends and Cycles of Ethereum

Debesh Bhowmik

8.1 INTRODUCTION

Vitalik Buterin a renowned programmer, invented Ethereum in 2013 although Gavin Wood, Charles Hoskinson, Anthony Di Iorio, and Joseph Lubin were also cofounders. Blockchain technology with a digital ledger promoted P2P exchanges for the advancement of smart contracts. During the forthcoming years, Ethereum was upgraded three or four times, which was called "phases" and known as Ethereum 2.0, which indicated that the network's consensus mechanism to proof of stake and to scale the network's transaction throughput with execution sharding would be improved along with EVM architecture.

Ethereum is considered the second-largest influential cryptocurrency in the world, followed by Bitcoin, and exchanged as digital tokens or coins under decentralized applications. The working of Ethereum is a mechanism of decentralized finance (DeFi) where digital ledgers and blockchains allow people to lend or borrow funds and earn interest in a savings-like account without the need for traditional intermediaries such as banks or non-banks. Gradually, it can be understood that Ethereum's influence has grown significantly, even more in 2020, after the Covid-19 outbreak.

The product of the current price and the total number of coins or tokens in circulation of Ethereum is called the market capitalization of Ethereum. On November 9, 2021, the price of Ethereum was \$4815, when Ethereum (Ether) had the highest market capitalization of \$571.67 billion in comparison with \$80 million on August 8, 2015, followed by the \$1 billion on March 12, 2016, which catapulted to \$500 billion on October 21, 2021, respectively. The market of Ethereum is cyclical, and after reaching its peak on November 9, 2021, the market capitalization has been falling downward in a cyclical fashion, although the upward cycle existed during the Covid-19 period (Figure 8.1).

Considering the above data, the estimation suggested that a 1% increase in the price of Ethereum per day leads to a 1.0618%t increase in the market capitalization of Ethereum per day during 07/8/2015 to 30/10/2023, which was found to be statistically significant at the 5% level.

$$\text{Log}(x) = 4.3008 + 1.0618\log(y)$$
$$(1448.74)^* (2119.23)^*$$

DOI: 10.1201/9781003453109-8

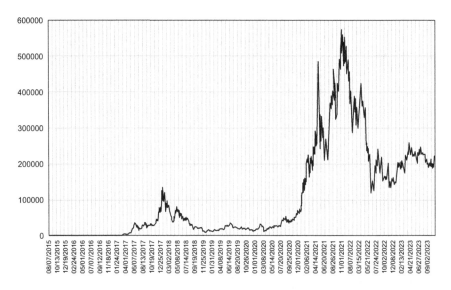

FIGURE 8.1 Market capitalization of Ethereum. Source: Plotted by the author with the help of data from https://etherscan.io/chart/marketcap

$R^2 = 0.999$, $F = 4491141*$, DW $= 0.0039$, $n = 3007$, $x =$ price of Ethereum, $y =$ value of market capitalization, $* =$ significant at the 5% level. A small value of DW implies that there is an autocorrelation problem. However, the demand for Ethereum is highly elastic.

It is astonishing that the supply of Ethereum has been unprecedentedly increasing and crossed 115 million in 2022, and then it started at a higher rate. The per-day issue of Ethereum showed a cyclical downward and after October 2020, it is L shaped. It implies that volatility reduced and virtual transactions after Covid-19 have fallen drastically (Figure 8.2).

8.2 ETHEREUM IS A DIGITAL ASSET

Whether cryptocurrency is to be considered as currency or asset or quasi-asset and on what conditions cryptocurrency will be considered as a stable coin digitally will evolve a long debate on monetary economics. The properties of an asset or quasi-asset are not identical with the properties of Ethereum as currency. It is argued that Ethereum is a digital asset that can be identifiable and its value realized. It is intangible and sellable online where the main value proposition is derived from its programmability through the Ethereum network. This digital asset is speculative and highly volatile with high risk. Ethereum is considered the Triple-Point Asset, constituting a store of value, capital assets, and consumable assets. It can be treated as a capital asset when it is staked because it generates yield and can therefore be valued based

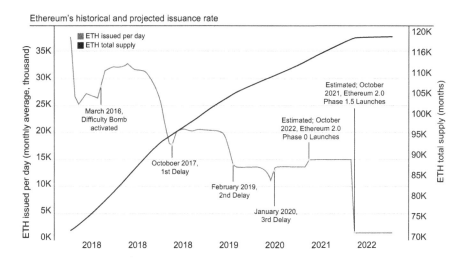

FIGURE 8.2 Issuance rate of Ethereum. Source: EthHub Activate, CoinDesk Research.

on its expected returns like as bonds. Economically, the staking yield is considerably higher than the interest earned on bonds. Ethereum generally acts as a store of value when holders deposit it to the decentralized finance protocols. Moreover, it takes on the role of a consumable asset when and how U.S. dollars are used to pay taxes. According to Kuper and Neureuter (2022), it is understood that a single true use case of Ethereum is to function as the naïve currency for transactions on the Ethereum blockchain. This type of function can be dramatically altered upon a successful transition to proof of stake.

But this digital asset is not real fiat currency because it is not the coin and paper money which is legal tender and is not digitally issued by a government's central bank. It is referred to as convertible virtual currency which can be digitally traded between users, and exchanged for or into real currencies or digital assets. All the following transactions are subject to taxation on the taxable income in cases of (i) sale of a digital asset for fiat, (ii) exchange of a digital asset for property, goods, or services, (iii) exchange or trade of one digital asset for another digital asset, (v) receipt of a digital asset as payment for goods or services, (v) receipt of a new digital asset as a result of a hard fork, (vi) receipt of a new digital asset as a result of mining or staking activities, and (vii) receipt of a digital asset as a result of an airdrop.

8.3   ETHEREUM IS NOT MONEY

Ethereum is a decentralized application protocol under the mechanism of exchange which empowers the decentralized finance and NFT movements. It will never be considered global legal tender money. The world free market will not converge on

Ethereum as money due to an undefined monetary policy, poor fiscal policy and higher centralization. There is no possibility of creating an alternate monetary system that can treat Ethereum as money. It is considered as the money of the Ethereum-based DeFi/NFT network (Santos-Alborna, 2021).

In spite of the acceptance of Ethereum as decentralized finance and its exchange-ability with US$ and some other currencies of the world due to agreement on government protocol abiding by taxable laws, Ethereum is still not widely accepted as a means of payment, store of value, and unit of account, and so it is not legal tender money or currency.

8.4   LITERATURE STUDIED

European Banking Authority (2014) examined more than seventy risks of using virtual currencies and found that they constitute substantial volatility, significant medium- and long-term risks of investing, risk of financial integration, money laundering, illegal financing, terrorism, and other illicit activities in which there are significant disadvantages of cryptocurrency. All these qualities have been preventing cryptocurrency from becoming a legitimate asset or currency.

Sovbetov (2018) examined the determinants that influence the prices of the most common five cryptocurrencies such as Bitcoin, Ethereum, Dash, Litecoin and Monero during 2010–2018 applying the ARDL technique using weekly data. The article explored that crypto-market-related factors such as market beta, trading volume, and volatility appeared as significant determinants for all five cryptocurrencies both in the short run and in the long run.

Antonakakis et al. (2019) used a Time-Varying Parameter Factor Augmented VAR model (TVP-FAVAR) to verify the transmission mechanism among nine leading cryptocurrencies. The authors found that Bitcoin is the most important transmitter of shocks in the cryptocurrency market, followed by Ethereum. The article found that connectedness measures among the nine cryptocurrencies exhibit large dynamic variability, with periods from high market uncertainty corresponding to strong connectedness and vice versa.

Mariana, Ekaputra, and Husodo (2021) found that both Bitcoin and Ethereum are suitable as short-term safe-havens during extreme stock market plunges. Moreover, Ethereum was plausibly a better safe-haven than Bitcoin during the pandemic. The article also examined that during the Covid-19 both Bitcoin and Ethereum returns were positively associated with gold return and negatively interrelated with stock return, respectively.

Shahzad et al. (2021) examined the daily return spillover among 18 cryptocurrencies during low- and high-volatility regimes by applying a Markov regime-switching (MS) vector autoregressive with exogenous variables (VARX) model. The article considered three pricing factors and the effect of the Covid-19 outbreak and found that the patterns of the spillover vary with time, which is consistent with contagion during stress periods. Its return spillover abruptly intensified during Covid-19, especially in the high volatility regime.

Harb, Bassil, Kassamani, and Baz (2022) identified an instantaneous positive bidirectional volatility spillover effect between Ethereum and Litecoin, and between Ethereum and Bitcoin.

The findings of the research by Havidz, Gaby, and Widjaja (2022) revealed that Ethereum price returns were greatly affected by Covid-19 factors, while macro-financial factors (stock indices and gold) had stronger effects on the return of Ethereum price rather than the crypto market.

Bhowmik (2022) analysed the role of Bitcoin and Ethereum in the economy through their trends during $2017m_{01}$–$2021m_{12}$ where the traditional regression model explained that the market capitalization of Bitcoin is positively related to the price of Bitcoin and the inflation rate, and negatively related to the price of Ethereum significantly. The cointegration and vector error correction model suggested that the market capitalization of Bitcoin has long-run causality with the price of Bitcoin, the price of Ethereum, and Ethereum and inflation rate. The cointegrating equation has been diverging away from equilibrium due to volatility. The price of Ethereum has short-run causality with the market capitalization of Bitcoin.

Li (2023) examined the relationship between Global Economic Policy Uncertainty (GEPU) and EthereumPrice by decomposing Ethereum into trend and cyclical components and extracting the cyclical component and excluding the irrelevant trend component during 2015–2022 and found that without the distortion from the trend component, GEPU Granger causes Ethereum. Additionally, GEPU and Ethereum cyclical patterns have a stable long-run negative cointegration relationship, which is positive and that may be disturbed by the trend component, of which the upward-sloping trend dominates. When Ethereum cyclical character does not have a trend, it responds negatively to GEPU and its lagging terms. This implies that cryptocurrencies may react ahead of the market, similar to stocks and other financial assets, when global economic uncertainty escalates. The error correction model demonstrates that the Ethereum cyclical component responds in the short run to changes in GEPU and deviations from their long-run equilibrium.

8.5   OBJECTIVE OF THE CHAPTER

The chapter endeavours to examine the behaviour of the Ethereum US Dollar rate, or in other words, the closing price of the Ethereum in terms of USD, from $2015m_{08}$ to $2023m_{05}$ which enabled it to show the nonlinear trend, cyclical trends, cycles, and seasonal fluctuations. Additionally, the chapter explores the AR and MA characteristics of the series and finally identifies the prediction trend of ARIMA(p,d,q) for $2025m_{01}$ which was also tested in the decomposed model of the Hamilton filter during the same period.

8.6   DATA AND METHODOLOGY

The nonlinear trend was fitted by the semi-log regression model.

The estimated equation can be written as: $\log(x_i) = a + bt + ct^2 + dt^3 + et^4 + ft^5 + ui$ where xi = variable to be estimated; $a, b, c, d, e,$ and f are constants, t = time (year), ui = random error, for all values of $i = 1, 2, 3, ..., n$.

Box and Jenkins' (1976) methodology of ARIMA (p, d, q) can be estimated as below.

$xt = a + bixt\_i + \varepsilon_t + b_0 i\varepsilon t\_i + \grave{e}t$ where x_t is the variable, a is a constant, bi are the coefficients of the AR process and boi are the coefficients of the MA process and $\grave{e}_t$ is the residual, and $i = 1, 2, ..., n$, and t = time. If bi and boi are less than zero and significant at the 5% level then the model is convergent and significant. If the roots of the AR and MA are less than one, then the model is stable and stationary.

Ljung and Box's (1978) Q-statistic is calculated as:

$$Q = T(T + 2) \, \Sigma r^2 k/(T - k) \text{ where } k = 1 \text{ to } s$$

Autocorrelation Function (ACF) can be derived from the formula:

$$\text{ACF} = \rho_s = a_1 \rho s_{-1} + a_2 \rho s_{-2} \qquad \text{where } s = 1, 2, 3, ...$$

And the Partial Autocorrelation Function (PACF) can be derived from the formula:

$$\Phi ss = (\rho_s - \Sigma \, \phi s_{-1}, \, \rho_{s-j})/(1 - \Sigma \, \phi_{s-1}, \, \rho_j) \text{ where } s = 3, 4, 5, ..., \phi sj = \phi_{s-1},$$
$$j - \phi_{ss}\phi_{s-1}, \text{s\_j}, j = 1, 2, 3, ...s - 1.$$

Hamilton (2018) regression filter for decomposition was applied to get cycles, cyclical trend, and seasonal variation utilising the STL method. In brief, the model can be stated as follows.

$yt_{+8} = \alpha_0 + \alpha_1 yt + \alpha_2 yt_{-1} + \alpha_3 yt_{-2} + \alpha_4 yt_{-3} + vt_{+8}$ where y = variable to be regressed.

Or, $vt_{+8} = yt_{+8} - (\grave{a}_0 + \grave{a}_1 yt + \grave{a}_2 yt_{-1} + \grave{a}_3 yt_{-2} + \grave{a}_4 yt_{-3})$

So, $yt = \alpha_0 + \alpha_1 yt_{-8} + \alpha_2 yt_{-9} + \alpha_3 yt_{-10} + \alpha_4 yt_{-11} + vt$

Therefore, $vt = y_t - (\grave{a}_0 + \grave{a}_1 yt_{-8} + \grave{a}_2 yt_{-9} + \grave{a}_3 yt_{-10} + \grave{a}_4 yt_{-11})$ where $\grave{a}_i$ is estimated.

$vt_{+}h = yt_{+}h - y_t$ is the difference i.e. how the series changes over h periods. For $h = 8$, the filter $1 - L^h$ wipes out any cycle with frequencies of exactly one year, thus taking out both the long-run trend as well as any strictly seasonal components.

It also applies to random walk: $y_t = yt_{-1} - \varepsilon t$ where $d = 1$ and $\omega_t^h = \varepsilon t_{+}h + \varepsilon t_{+}h_{-1} + ... + \varepsilon t_{+1}$

Regression filter reduces to a difference filter when applied to a random walk. Hamilton suggested $h = 8$ for business cycles and $h = 20$ for studies in financial cycles. Regression vt converges in large samples to $\alpha i = 1$ and all other $\alpha_j = 0$. Thus, the forecast error is $vt_{+}h = yt_{+}h - y_t$.

The residual equation vt can be decomposed into trend, cycle, and seasonally adjusted components through SEATS/TRAMO or STL or census X-13 packages. The STL method was developed by Cleveland, Cleveland, McRae, and Terpenning (1990).

The monthly data on the closing price of the Ethereum US$ rate were collected from https://www.marketwatch.com/investing/cryptocurrency/ethusd/download =data?startdate=8/31/20215.The daily data on the market capitalization of Ethereum was collected from https://etherscan.io/chart/marketcap

8.7 THE OBSERVATIONS ON THE TRENDS AND PATTERNS OF ETHEREUM

The closing price of Ethereum in terms of US$ during $2015m_{08} - 2023m_{05}$ constitutes a cyclical pattern of being peaks and troughs. From the value of 2.2 in $2015m_{12}$ it started to increase up to 8.07 in $2016m_{12}$ and reached a peak at 1105 in $2018m_1$ and fell down to 152 in $2019m_{11}$, then began to catapult to a top of 4829 in $2021m_{11}$ and started to decline to 1199 in $2022m_{12}$ then moved to an upswing. As a whole, there is upward trend up to 2023, although the highest upward trend with the peak was observed during the Covid-19 period and after that, it is declining. From $2015m_{08}$ to $2023m_{05}$, the trend of Ethereum price is seen as strongly volatile, which is depicted in Figure 8.3.

The long-run trend of the closing price of Ethereum from $2015m_{08}$ to $2023m_{05}$ has four phases of upward and downward trends that are statistically significant at the

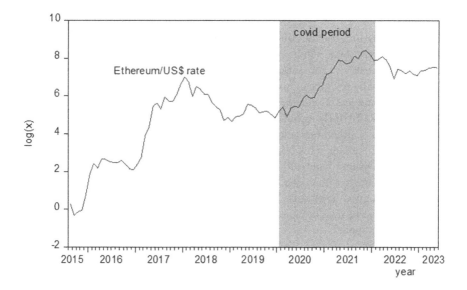

FIGURE 8.3 Price of Ethereum. Source: Plotted by the author.

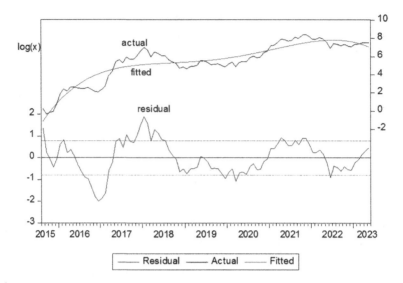

FIGURE 8.4   The trend of Ethereum closing price. Source: Plotted by the author.

5% level. The estimated trend is given below. It has strong autocorrelation problem. In Figure 8.4 the trend is shown clearly.

$$\text{Log}(x) = -1.681 + 0.587t - 0.0186t^2 + 0.000259t^3 - 1.2e^{\,06}t^4 + ui$$
$$(-3.83) * (9.29) * (-6.94) * (6.11) *\quad (-5.58)*$$

$R^2 = 0.87$, $F = 151.04*$, AIC $= 2.43$, SC $= 2.56$, DW $= 0.21$, $n = 94,*$ = significant at the 5% level.

It has a break unit root at $2016m_{12}$ where the ADF test statistic $(-3.29\ (p = 0.50))$ is accepted at $H_0 =$ it has a unit root at the 5% significant level, but its first difference series has no break unit root whose ADF $= -9.11\ (p < 0.01)$ at $H_0 =$ has unit root has been automatically rejected. This is shown in Figure 8.5.

For the automatically selected ARIMA(3,1,2) model using the ARMA maximum likelihood (OPG-BHHH) technique for the closing price of Ethereum, we have estimated equation of Ethereum's closing price which is moving towards equilibrium in an insignificant manner. This is because the t-values of the coefficients AR and MA are found to be insignificant, although the values of the coefficients are less than one and their roots are less than one, indicating that the model is convergent and stable. The model is a good fit because the AIC and SC values are minimum, R^2 is maximum and t-value of the coefficient of σ^2 is significant, which implies its minimum volatility in this model.

$$d\log(x) = 0.0758 + 0.129d\log(x)t_{-3} + 0.0933\varepsilon\,t_{-2} + 0.125\sigma^2 t$$
$$(1.50)\qquad (1.11)\qquad\quad (0.88)\qquad\quad (7.20)*$$

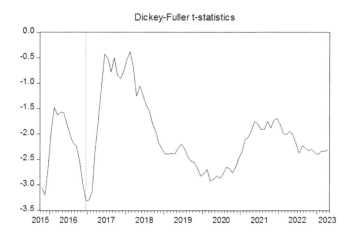

FIGURE 8.5 Break unit root. Source: Plotted by the author.

$R^2 = 0.028$, $F = 0.86$, DW = 1.62, AIC = 0.84, SC = 0.95, $n = 93$, AR roots = 0.51, $-0.25 \pm 0.44i$, MA roots = $-0.00 \pm 0.31i$,* = significant at the 5% level.

In Figure 8.6, the fitted line has been approaching zero but has not reached since *t*-values are not significant, and the actual values containing huge volatilities have crossed many times with the fitted values. The fitted ARIMA(3,1,2) is cyclical.

Although the model of the closing price of the Ethereum has been approaching equilibrium in a convergent manner, its volatility has not disappeared completely. This can be represented by the AC and PAC behaviour of this ARIMA(3,1,2) model because the movements of AC and PAC vary from negative to positive values

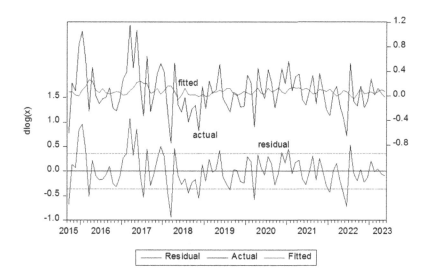

FIGURE 8.6 ARIMA(3,1,2). Source: Plotted by the author.

continuously, although their values are significant since they pass through the $\pm 5\%$ significant level, as shown in Figure 8.7.

The ARIMA(3,1,2) model of the closing price of the Ethereum has been forecasted for the forthcoming period $2025m_{01}$ which implies that the forecasting path is turned into the seasonal fluctuations in the "V" shape that has moved away from equilibrium and shows increased volatility. This is plotted in Figure 8.8 . It implies that the closing price of Ethereum has the character of being nonstationary and volatile in the forthcoming periods.

Now, the decomposition of trends, cycles, and seasonal variation of the closing price of the Ethereum during $2015m_{08} - 2023m_{05}$ can be found out by using the Hamilton regression filter model which is estimated below.

$$\text{Log}(x) = 5.714 + 0.174 \log(x)t_{-24} - 0.051 \log(x)t_{-25} - 0.202 \log(x)t_{-26} + 0.244$$
$$\log(x)t_{-27} + vt$$

| (13.5)* | (0.44) | (−0.08) | (−0.34) | (0.65) |

Autocorrelation	Partial Correlation		AC	PAC	Q-Stat	Prob
		1	0.166	0.166	2.6308	
		2	-0.008	-0.037	2.6374	
		3	0.006	0.014	2.6409	0.104
		4	-0.078	-0.084	3.2506	0.197
		5	0.002	0.032	3.2512	0.354
		6	-0.042	-0.054	3.4267	0.489
		7	-0.050	-0.031	3.6806	0.596
		8	0.122	0.131	5.2189	0.516
		9	-0.005	-0.050	5.2219	0.633
		10	-0.169	-0.168	8.2455	0.410
		11	0.016	0.071	8.2728	0.507
		12	-0.120	-0.131	9.8434	0.454
		13	0.037	0.083	9.9952	0.531
		14	0.050	0.007	10.279	0.592
		15	0.101	0.132	11.441	0.574
		16	0.154	0.066	14.174	0.437
		17	-0.137	-0.197	16.359	0.359
		18	-0.138	-0.039	18.609	0.289
		19	-0.096	-0.098	19.702	0.290
		20	-0.132	-0.092	21.826	0.240
		21	0.052	0.109	22.159	0.276
		22	0.007	-0.071	22.165	0.332
		23	0.006	0.043	22.170	0.390
		24	0.177	0.111	26.163	0.245
		25	-0.010	0.004	26.176	0.293
		26	-0.105	-0.074	27.641	0.275
		27	-0.054	-0.088	28.025	0.307
		28	-0.146	-0.098	30.935	0.231
		29	-0.055	-0.135	31.354	0.257
		30	0.063	0.055	31.911	0.278
		31	-0.159	-0.181	35.516	0.188
		32	-0.007	0.008	35.524	0.224
		33	-0.065	0.012	36.149	0.241
		34	-0.084	-0.014	37.213	0.241
		35	0.016	0.020	37.252	0.280
		36	0.007	-0.003	37.260	0.321

FIGURE 8.7   AC PAC of the ARIMA(3,1,2) model. Source: Plotted by the author.

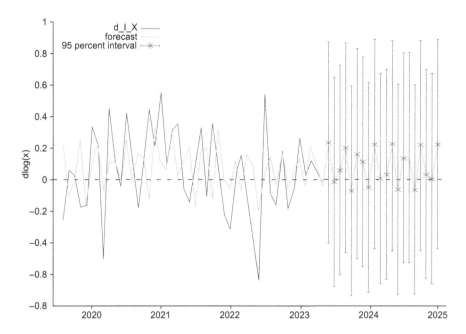

FIGURE 8.8   Forecast ARIMA.

$R^2 = 0.06$, $F = 1.14$, AIC = 3.17, SC = 3.34, DW = 0.07, $n = 67$,* = significant at the 5% level.

Thus, the residual vt series has been obtained from the estimated equation, which is given below.

$$Vt = \log(x) - [\ 5.714 + 0.174\ \log(x)t_{-24} - 0.051\ \log(x)t_{-25} - 0.202\ \log(x)t_{-26} + 0.244$$
$$\log(x)t_{-27}]$$

This vt has been decomposed by using the STL method that constitutes cycles, cyclical trends, and seasonal variation of the closing price of the Ethereum price during $2015m_{08} - 2023m_{05}$. In panel 1 of Figure 8.9, the cycle shows 17 peaks and 17 troughs. In panel 2, the cyclical trend clearly shows 2 peaks and one trough where the trough takes longer time. In panel 3,there are huge peaks and troughs in seasonal fluctuations, and their amplitudes are short.

The vt has a break unit root in which ADF = -2.849 whose probability is 0.7629. This is accepted for the break unit root at $H_0 =$ it has a unit root when applying the break unit root test. The point $2020m_{10}$ has been detected as a break unit root point, which is shown in Figure 8.10.

With usual reason, the first difference series of vt has no unit root since ADF = -8.293 whose probability becomes $p < 0.01$ which is rejected for $H_0 =$ it has a unit root.

Now, the automatically selected ARIMA(2,0,1) model of vt has been estimated below as:

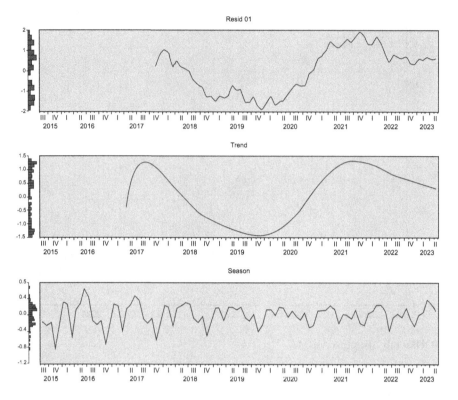

FIGURE 8.9 Decomposition of closing price of Ethereum. Source: Plotted by the author.

$$Vt = 0.1497 + 0.906 \, vt_{-2} + 0.992\varepsilon \, t_{-1} + 0.0864\sigma^2 t$$
$$(0.18) \quad (14.58)^* \quad (10.32)^* \quad (4.32)^*$$

$R^2 = 0.92$, $F = 272.76^*$, AIC = 0.55, SC = 0.68, DW = 1.87, $n = 67$,* = significant at the 5% level. Inverted AR roots = $\pm$ 0.95, Inverted MA root = -0.99.

Where t-values of the coefficients of AR and MA have been found significant at the 5% level and the coefficients are less than one, even AR roots and MA roots are less than one, therefore ARIMA(2,0,1) is convergent and stable. Moreover, AIC is minimum, R^2 is maximum, and the t-value of the coefficient of σ^2 is significant, which implies that it showed minimum volatility. Thus, the model is approaching equilibrium significantly, as depicted in Figure 8.11.

The residual test of ARIMA(2,0,1) in the correlogram verified that its AC and PAC move from positive to negative values, proving its seasonal variation although their limits pass through the $\pm$5% significant level and they are not diminishing, as shown in Figure 8.12. It implies that the closing price of the Ethereum during $2015m_{08} - 2023m_{05}$ suffers from seasonal variation even if it passes through significant ARIMA(2,0,1).

If this ARIMA(2,0,1) of v_t has been forecasted for the forthcoming period of $2025m_{01}$ then the forecast path has been converging towards equilibrium significantly

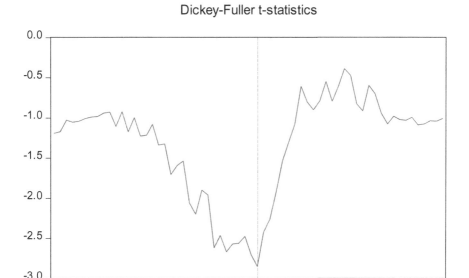

FIGURE 8.10 Break unit root. Source: Plotted by the author.

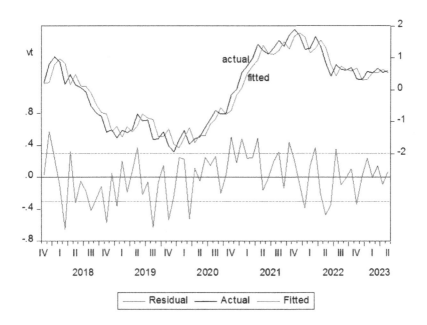

FIGURE 8.11 Fitted ARIMA. Source: Plotted by the author.

Autocorrelation	Partial Correlation		AC	PAC	Q-Stat	Prob
		1	0.061	0.061	0.2648	
		2	0.025	0.021	0.3092	
		3	0.096	0.093	0.9712	0.324
		4	0.229	0.220	4.8287	0.089
		5	0.145	0.127	6.3905	0.094
		6	0.012	-0.011	6.4022	0.171
		7	0.073	0.033	6.8145	0.235
		8	0.038	-0.039	6.9278	0.328
		9	0.008	-0.057	6.9326	0.436
		10	0.014	-0.013	6.9487	0.542
		11	-0.016	-0.042	6.9689	0.640
		12	0.046	0.039	7.1452	0.712
		13	-0.082	-0.078	7.7252	0.738
		14	-0.037	-0.029	7.8455	0.797
		15	0.028	0.041	7.9166	0.849
		16	-0.019	-0.015	7.9494	0.892
		17	-0.136	-0.114	9.6498	0.841
		18	0.082	0.135	10.277	0.852
		19	-0.191	-0.227	13.776	0.683
		20	-0.085	-0.053	14.479	0.697
		21	0.051	0.125	14.735	0.739
		22	-0.061	-0.081	15.114	0.770
		23	-0.200	-0.150	19.325	0.564
		24	-0.156	-0.059	21.948	0.463
		25	0.052	0.047	22.251	0.505
		26	0.012	0.058	22.268	0.563
		27	-0.229	-0.162	28.327	0.293
		28	-0.156	-0.097	31.226	0.220

FIGURE 8.12   AC and PAC of ARIMA(2,0,1) of vt. Source: Plotted by the author.

without fluctuations in the forecast path of the closing price of the Ethereum which is depicted in Figure 8.13.

8.8 LIMITATIONS

This chapter has its limitations too. The ARCH test is absent here. Even the behaviour of structural breaks was not analysed since unit break root test has been included. The chapter can be extended by explaining the determinants of the price of Ethereum and its market capitalization which could explain the capital market behaviour in detail.

8.9 POLICY IMPLICATIONS

To regulate Ethereum and cryptocurrencies, its ban is important; the US government is collaborating with other countries, but IMF proposes that countries, including

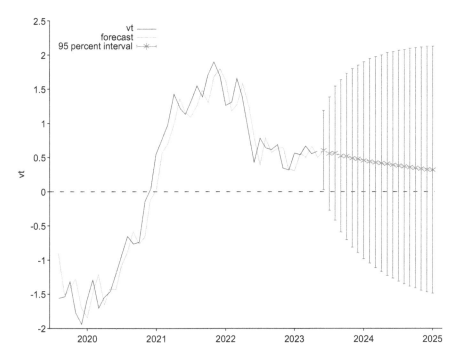

FIGURE 8.13   Forecast of ARIMA(2,0,1) of v_t. Source: Plotted by the author.

India, should not ban cryptocurrencies completely. Instead, it suggests introducing taxation as a percentage of transactions. The IMF, Federal Reserve Bank, Reserve Bank of India, and other central banks plan to introduce central bank digital currency as legal tender to counter the accelerated market capitalization of cryptos. When the price of Ethereum is cyclically increasing, the government can regulate the exchange rate by introducing various ranges and can hike transaction fees according to these ranges. A team of digital scams should be formed for cybercrime and fraud transactions. I think the stable coin theory is ineffective, so that government supervision on crypto exchanges should be the proper step. Verification of accounts where cryptos have been converted to US dollars or other liquid assets is urgently needed.

8.10   CONCLUSION

The chapter finally verified that the trend of the closing price of Ethereum in terms of USD, from $2015m_{08}$ to $2023m_{05}$ is nonlinear, having upward and downward cycles, which showed break unit root at $2016m_{12}$ and automatically selected ARIMA(3,1,2) model is convergent and stable, moves to equilibrium insignificantly, and its forecast path for $2025m_{01}$ is seasonally fluctuated. On the otherhand, its Hamilton filter decomposition constitutes huge peaks and troughs in cycles, with two peaks and one trough in the cyclical trend and V-shaped seasonal fluctuations. If the residual of the Hamilton regression filter for the closing price of the Ethereum passes through the automatically selected ARIMA(2,0,1) model, it tends to equilibrium in a stable and convergent manner. However, both AC and PAC exist during seasonal variation, but its predicted forecast path for $2025m_{01}$ has the character of significant convergence towards equilibrium without seasonal fluctuations.

REFERENCES

Antonakakis, N., Chatziantoniou, I., & Gabauer, D. (2019). Cryptocurrency market contagion: Market uncertainty, market complexity, and dynamic portfolios. *Journal of International Financial Markets, Institutions and Money, 61*, 37–51. https://doi.org/10 .1016/J.INTFIN.2019.02.003

Bhowmik, D. (2022, February). Crypto-Currency: Trends and Determinants. *Saudi Journal of Economics and Finance, 6*(2), 37–50. -https://doi.org/10.36348/sjef.2022.v06i02.001

Box, G., & Jenkins, G. (1976). *Time Series Analysis, Forecasting and Control*. San Francisco: Holden Day. http://garfield.library.upenn.edu/classics1989/A1989AV48500001.pdf

Cleveland, R. B., Cleveland, W. S., McRae, J. E., & Terpenning, I. (1990). STL: A seasonal-trend decomposition. *Journal of Official Statistics, 6*, 3–73. https://www.wessa.net/ download/stl.pdf

European Banking Authority. (2014). Eba opinion on virtual currencies. https://www .eba.europa.eu/documents/10180/657547/EBA-Op-2014-08+Opinion+on+Virtual +Currencies.pdf

Hamilton, J. D. (2018). Why you should never use the Hodrick Prescott Filter. *Review of Economics and Statistics, 100*(5), 831–843. https://ideas.repec.org/a/tpr/restat/ v100y2018i5p831-843.html

Harb, E., Bassil, C., Kassamani, T., & Baz, R. (2022). Volatility interdependence between cryptocurrencies, equity, and bond markets. *Computational Economics*. https://link .springer.com/article/10.1007/s10614-022-10318-7#citeas

Havidz, S. A. H., Gaby, T., & Widjaja, M. (2022, April). Is Ethereum's Price Return Determined by COVID-19, Macro-financial, and Crypto Market Factors? ICEEG '22: Proceedings of the 6th International Conference on E-Commerce, E-Business and E-Government, 237–242. https://doi.org/10.1145/3537693.3537714

Kuper, C., & Neureuter, J. (2022, May). Ethereum: Advancing Blockchain with Smart Contracts and Decentralized Applications. www.fidalitydigital.com

Li, M. (2023). Global Economic Policy Uncertainty and Ethereum Price-A Time-Series Analysis from 2015 to 2022. https://www.atlantis-press.com/article/125982679.pdf

Ljung, G., & Box, G. (1978). On a measure of Lack of Fit in Time Series Models. *Biometrica, 65*, 297–303. https://doi.org/10.1093/biomet/65.2.297

Mariana, C. D., Ekaputra, I. A., & Husodo, Z. A. (2021, January). Are Bitcoin and Ethereum Safe-havens for Stocks during the COVID-19 Pandemic? *National Library of Medicine*. https://www.ncbi.nlm.nih.gov/pmc/articles/PMC7566681/

Santos-Alborna, A. (2021, November). Ethereum Is Not Money and That's Okay. https://seek-ingalpha.com/article/4471644-ethereum-not-money-that-is-okay

Shahzad, S. J. H., Bouri, E., Kang, S. H., & Saeed, T. (2021). Regime specific spillover across cryptocurrencies and the role of COVID-19. *Financial Innovation*. https://doi.org/10 .1186/s40854-020-00210-4

Sovbetov, Y. (2018). Factors Influencing Cryptocurrency Prices: Evidence from Bitcoin, Ethereum, Dash, Litcoin, and Monero. *Journal of Economics and Financial Analysis, 2*(2), 1–27. http://dx.doi.org/10.1991/jefa.v2i2.a16

9 The Future of Ledger Development and Smart Contracts

*Minal Shukla, Biswaranjan Acharya,
and Asik Rahaman Jamader*

9.1 INTRODUCTION

In the ever-evolving landscape of blockchain technology, the future of ledger development and smart contracts promises to be nothing short of revolutionary. As we stand on the precipice of a new era, the convergence of distributed ledger technology and self-executing contracts is reshaping industries, redefining trust, and unlocking unprecedented possibilities for businesses and individuals alike (Benahmed et al., 2019).

The term "ledger development" might sound like an arcane subject, but it lies at the heart of blockchain innovation. Ledgers are no longer confined to a simple record of financial transactions; they have grown into a distributed, tamper-resistant, and transparent foundation for a wide array of applications. These applications are not only transforming the financial sector but also extending their reach into supply chains, healthcare, real estate, and beyond. In this exciting era, the future of ledger development takes center stage, offering a glimpse into a world where information is not just stored but secured, shared, and harnessed with unparalleled efficiency (Das et al., 2022).

At the heart of this transformation are smart contracts. These self-executing, code-based agreements have the potential to eliminate intermediaries, automate complex processes, and ensure trust without relying on centralized authorities (Barman, A., Barman, M., & Jamader, A. R., 2024). In the world of blockchain, smart contracts are the catalysts for change, empowering individuals and organizations to streamline their operations, reduce costs, and enhance security.

This exploration into the future of ledger development and smart contracts invites us to delve into the fascinating possibilities and challenges that lie ahead. We will journey through the evolving blockchain landscape, uncovering the latest innovations and breakthroughs that promise to redefine industries and disrupt established norms. With each section, we will unravel the potential applications of blockchain technology and the profound impact it could have on our daily lives (Das et al., 2019).

DOI: 10.1201/9781003453109-9

As we embark on this journey, it is crucial to recognize that the future of ledger development and smart contracts is not a single, predetermined path but a continually branching road, shaped by the creativity and ingenuity of developers, entrepreneurs, and visionaries. The blockchain landscape is teeming with opportunities, and the future we are about to explore is limited only by our imagination and innovation (Jamader et al., 2023).

Join us as we navigate through the exciting terrain of ledger development and smart contracts, where the possibilities are as limitless as the technology itself, and where the promise of a more decentralized, efficient, and secure world awaits us (Juels, A., Kosba, A., & Shi, E., 2016).

9.2 HISTORICAL CONTEXT

To fully appreciate the future of ledger development and smart contracts, it's imperative to understand the historical context that has paved the way for these transformative technologies (Jamader et al., 2021). The roots of blockchain and smart contracts trace back to the late 20th century, and the journey from their inception to their current state of rapid evolution is a testament to the persistent pursuit of more efficient, secure, and decentralized systems.

9.2.1 THE BIRTH OF CRYPTOGRAPHY

The historical backdrop for ledger development and smart contracts begins with the evolution of cryptography, particularly public-key cryptography in the 1970s. These innovations laid the groundwork for secure, digital communication, enabling the creation of encrypted transactions and secure data storage, both critical elements of blockchain technology (Jamader et al., 2023)

1. The Cypherpunk Movement: In the late 1980s and early 1990s, the cypherpunk movement emerged. This group of cryptographers, mathematicians, and privacy advocates aimed to use cryptography to protect individual privacy and ensure freedom in the digital age. They explored concepts that were foundational to blockchain, such as digital cash and secure communications, emphasizing the need for decentralized systems to counter centralized control.
2. Bitcoin and the Genesis of Blockchain: The watershed moment in the history of blockchain and ledger development came with the release of the Bitcoin whitepaper in 2008 by an anonymous entity known as Satoshi Nakamoto. Bitcoin introduced a groundbreaking decentralized ledger, the blockchain, which enabled peer-to-peer transactions without the need for intermediaries. It marked a shift from traditional financial systems to a trustless, transparent, and tamper-resistant ledger system.
3. Ethereum and Smart Contracts: In 2015, Ethereum, led by Vitalik Buterin, took the blockchain concept a step further with the introduction of smart contracts. These self-executing contracts, written in code, allowed for

automated and trustless agreements. Ethereum's introduction of a generalized blockchain platform for executing smart contracts expanded the potential of blockchain technology beyond just cryptocurrency.

4. The Proliferation of Blockchain Applications: Over the years, blockchain technology has found applications far beyond its initial purpose of digital currencies. It has been adopted in supply chain management, healthcare, voting systems, real estate, and many other sectors, showcasing its versatility and potential to revolutionize various industries.

5. Challenges and Regulatory Developments: As blockchain and smart contracts gained traction, they also encountered challenges related to scalability, security, and regulatory concerns. Governments and regulatory bodies started to take an interest in this technology, leading to a mix of regulatory approaches globally.

6. Ongoing Innovation and Standardization: The historical context for the future of ledger development and smart contracts includes a backdrop of constant innovation. Development teams, businesses, and organizations worldwide are actively working on improving the technology, addressing its limitations, and creating standardized protocols to facilitate broader adoption (Jamader et al., 2019).

Understanding this historical context sets the stage for exploring the promising developments and challenges that lie ahead in the world of ledger development and smart contracts. As we move forward, it is clear that the journey from the inception of blockchain technology to its current state is a testament to human innovation and our unyielding pursuit of a more efficient, secure, and decentralized digital world.

9.3 EMERGING TRENDS AND INNOVATIONS

The landscape of ledger development and smart contracts is continually evolving, driven by a relentless wave of innovation and adaptation. In this chapter, we will explore the most promising emerging trends and innovations that are shaping the future of these technologies.

9.3.1 LAYER-2 SCALING SOLUTIONS

Scalability issues have been a persistent concern for blockchain networks. Emerging layer-2 scaling solutions, such as state channels and sidechains, are gaining traction. These solutions aim to alleviate the congestion and high transaction costs on the main blockchain by processing most transactions off-chain while ensuring security and trust through periodic settlements (Jamader, 2022).

9.3.2 CROSS-CHAIN INTEROPERABILITY

Interoperability between different blockchains is becoming a central focus. Projects like Polkadot and Cosmos are pioneering cross-chain solutions, enabling seamless communication and asset transfers between blockchains. These initiatives have the potential to unlock new use cases and extend the reach of blockchain technology.

9.3.3   DeFi Evolution

Decentralized Finance (DeFi) continues to evolve and expand. New DeFi protocols are being developed to address the limitations of existing platforms, with an emphasis on security, user experience, and cross-platform compatibility. Innovations in stablecoins, decentralized exchanges, and yield farming are driving the growth of the DeFi ecosystem.

9.3.4   Non-Fungible Tokens (NFTs) Beyond Art

NFTs have transcended their origins in the art world. They are now being used in various industries, including gaming, music, virtual real estate, and collectibles. Innovations in NFT standards and marketplaces are making it easier for creators to tokenize and monetize their digital assets.

9.3.5   Hybrid and Multichain Solutions

Hybrid blockchain networks that combine aspects of both public and private blockchains are gaining attention, offering the security and decentralization of public networks with the privacy and control of private networks. Multi-chain ecosystems, like those enabled by Polkadot, aim to create a web of interconnected blockchains with specialized functions.

9.3.6   Blockchain as a Service (BaaS)

Blockchain as service platforms, provided by major cloud service providers, simplify the deployment and management of blockchain networks. This trend is making blockchain technology more accessible to enterprises and startups, facilitating rapid development and deployment of applications.

9.3.7   Enhanced Smart Contract Languages

Smart contract languages are evolving to provide more security and flexibility. Innovations include formal verification tools that rigorously test smart contract code for vulnerabilities, and the development of higher-level languages that simplify contract creation and reduce human errors.

9.3.8   Decentralized Autonomous Organizations (DAOs)

DAOs are evolving as self-governing entities, enabling decentralized decision-making and resource allocation. New DAO structures and methods for preventing manipulation and fraud are under exploration, potentially revolutionizing governance in various organizations.

9.3.9 QUANTUM-RESISTANT BLOCKCHAIN

With the emergence of quantum computing, the security of existing blockchain networks is at risk. Innovations in quantum-resistant cryptographic algorithms and the development of quantum-resistant blockchains are aimed at protecting the long-term viability of blockchain technology.

9.3.10 GREEN AND SUSTAINABLE BLOCKCHAIN SOLUTIONS

Efforts to reduce the environmental impact of blockchain networks are growing. New consensus mechanisms, like proof of stake and delegated proof of stake, are being adopted to reduce energy consumption. Some projects are even integrating blockchain technology with renewable energy solutions.

9.3.11 TOKENIZATION OF REAL-WORLD ASSETS

The tokenization of real-world assets, such as real estate, fine art, and company equity, is gaining traction. Innovations in regulatory frameworks and compliance solutions are making it easier to represent these assets as blockchain tokens, increasing accessibility and liquidity.

As we navigate these emerging trends and innovations, it is clear that the future of ledger development and smart contracts is a dynamic and ever-adapting landscape. These advancements offer a glimpse into the transformative potential of blockchain technology, emphasizing the ongoing need for creativity, collaboration, and responsible development to ensure a sustainable and inclusive future (Jamader, A. R., Das, P., & Acharya, B., 2022).

9.4 FUTURE PROSPECTS

The future of ledger development and smart contracts is filled with boundless potential and exciting prospects. In this section, we will explore the directions in which these technologies are heading, the transformative impacts they may have on various industries, and the broader implications for society in the years to come.

9.4.1 MAINSTREAM ADOPTION

Blockchain technology and smart contracts are on the cusp of widespread adoption. As user-friendly interfaces and scalable solutions continue to emerge, more individuals and businesses will integrate blockchain into their daily operations, further expanding its reach.

9.4.2 GLOBAL FINANCIAL INCLUSION

Blockchain and smart contracts have the potential to bring financial services to the unbanked and underbanked populations worldwide. Decentralized finance (DeFi) platforms and cross-border payment solutions could enable more equitable access to financial resources (Jamader et al., 2023).

9.4.3 IMPROVED SUPPLY CHAIN MANAGEMENT

Blockchain's transparency and traceability will revolutionize supply chain management, reducing fraud and ensuring the authenticity of products. Consumers will have unprecedented insights into the origin and journey of the products they purchase.

9.4.4 ENHANCED DATA SECURITY AND PRIVACY

Innovations in zero-knowledge proofs, privacy coins, and privacy-focused blockchains will bolster data security and privacy, giving individuals greater control over their personal information while meeting regulatory requirements.

9.4.5 HEALTHCARE REVOLUTION

Blockchain's role in securely managing healthcare data will enhance patient privacy and streamline medical record-sharing. Telemedicine and remote monitoring could be powered by smart contracts, improving the efficiency of healthcare services.

9.4.6 VOTING AND GOVERNANCE

Blockchain technology will make secure, transparent, and verifiable voting systems a reality, potentially increasing voter participation and trust in electoral processes. Blockchain-based governance systems will reshape the way organizations and communities make decisions.

9.4.7 ENERGY EFFICIENCY

The transition to energy-efficient consensus mechanisms, along with the integration of blockchain technology with renewable energy solutions, has the potential to reduce the environmental footprint of blockchain networks.

9.4.8 ART AND CREATIVITY

NFTs and blockchain technology will continue to transform the art and creative industries, empowering artists, musicians, and content creators with new monetization opportunities and creative control.

9.4.9 SECURE DIGITAL IDENTITY

Blockchain can provide a secure and self-sovereign digital identity for individuals, reducing identity theft and ensuring data ownership. Users will have control over who accesses their identity information.

9.4.10 DECENTRALIZED AUTONOMOUS ORGANIZATIONS (DAOS)

DAOs will evolve into self-sustaining entities, allowing for decentralized decision-making and resource allocation. They may challenge traditional corporate structures and introduce new forms of governance and economic organization.

9.4.11 QUANTUM-RESISTANT SECURITY

The development of quantum-resistant blockchain networks and cryptographic solutions will ensure the long-term security of digital assets and data.

9.4.12 TOKENIZATION OF REAL-WORLD ASSETS

Real-world assets, from real estate to company equity, will continue to be tokenized, increasing liquidity and enabling a broader range of investors to access traditionally illiquid markets (Kirillov et al., 2019).

9.4.13 SPACE AND BEYOND

Blockchain technology may play a role in resource allocation, governance, and secure communications in future space exploration and colonization efforts.

9.4.14 ETHICAL AND SOCIAL IMPACTS

The adoption of blockchain and smart contracts will raise ethical questions about privacy, governance, and digital rights. Society will need to navigate these issues to ensure responsible development and equitable access.

As we look to the future, the prospects for ledger development and smart contracts are breathtaking (Li & Kassem 2021). They offer a path to a more decentralized, efficient, secure, and inclusive world. However, it is essential to approach these prospects with responsibility, foresight, and a commitment to addressing the challenges and concerns that come with this transformative technology. In the chapters to come, we will explore these issues in greater depth, offering insights into how the future can be shaped for the better.

9.5 REGULATORY AND LEGAL CONSIDERATIONS

In the ever-evolving landscape of ledger development and smart contracts, regulatory and legal considerations play a pivotal role in shaping the future. This section delves into the multifaceted aspects of blockchain and smart contract regulations, outlining the challenges, solutions, and their profound impact on the technology's evolution.

9.5.1 EVOLVING REGULATORY FRAMEWORKS

The regulatory landscape for blockchain and smart contracts is a patchwork of different approaches worldwide. Governments and regulatory bodies are in the process of adapting existing legal frameworks or creating new ones to address the complexities of blockchain technology (Nayak et al., 2022).

9.5.2 AML AND KYC COMPLIANCE

Anti-Money Laundering (AML) and Know Your Customer (KYC) regulations are of particular concern, as blockchain's pseudonymity can potentially be exploited for illicit purposes. Regulatory compliance measures are being implemented to ensure transparency and security in blockchain-based transactions (Hamilton, 2020).

9.5.3 SECURITIES AND TOKENIZATION

The classification of tokens as securities is a significant concern. Regulators are determining which tokens fall under security regulations, with implications for Initial Coin Offerings (ICOs) and Security Token Offerings (STOs). Clarity in this area is crucial to prevent legal issues and protect investors.

9.5.4 SMART CONTRACT LEGAL ENFORCEABILITY

The legal enforceability of smart contracts is a complex issue. Traditional legal systems are still adapting to the concept of self-executing contracts, and ensuring that they are legally binding and recognized in court remains a challenge.

9.5.5 PRIVACY AND DATA PROTECTION

Data privacy regulations, such as the European General Data Protection Regulation (GDPR), have implications for blockchain networks that store personal information. Ensuring compliance while maintaining blockchain's transparency is a delicate balancing act.

9.5.6 DIGITAL IDENTITY AND AUTHENTICATION

Blockchain's role in digital identity management is raising questions about authentication and user control over personal data. Regulatory frameworks that protect individuals' digital identities and data are being developed.

9.5.7 CROSS-BORDER TRANSACTIONS

Cross-border transactions can be subject to multiple regulatory jurisdictions, creating complexity. Harmonizing international regulations for blockchain and cross-border transactions is essential to facilitate global trade and financial interactions.

9.5.8 TAXATION AND REPORTING

Blockchain-based transactions and assets pose challenges for taxation and reporting. Ensuring that individuals and entities meet tax obligations and report crypto-related income is a growing concern for tax authorities.

9.5.9 LEGAL DISPUTE RESOLUTION

Resolving legal disputes related to blockchain and smart contracts requires specialized knowledge. The legal community is adapting by developing expertise in blockchain law and arbitration mechanisms for blockchain disputes.

9.5.10 REGULATORY SANDBOXES

Many jurisdictions are establishing regulatory sandboxes to allow businesses and startups to experiment with blockchain technology within a controlled and supervised environment. These sandboxes promote innovation while ensuring compliance with regulatory requirements.

9.5.11 POLICY INNOVATION

Regulators are increasingly working in collaboration with the blockchain community to develop policies that foster innovation and protect the interests of both consumers and the industry. Such collaboration aims to strike a balance between regulatory oversight and technological advancement.

9.5.12 FUTURE-PROOFING REGULATIONS

Anticipating the pace of technological change in blockchain and smart contracts, regulators are increasingly focused on creating flexible regulatory frameworks that can adapt to evolving technology.

As blockchain and smart contracts continue to redefine industries and challenge traditional legal norms, this chapter explores the dynamic nature of blockchain regulation, highlighting both the challenges and opportunities. A balanced regulatory framework, coupled with legal innovation, will be essential to ensuring that blockchain and smart contracts can thrive while maintaining the necessary safeguards for consumers, businesses, and society at large (Khan et al., 2021).

9.6 INDUSTRY ADOPTION

The future of ledger development and smart contracts is intrinsically tied to their adoption across various industries. This section explores the current state and potential future of adoption, shedding light on how these technologies are reshaping businesses and sectors, and paving the way for innovation.

9.6.1 FINANCE AND BANKING

The financial sector has been an early and enthusiastic adopter of blockchain and smart contract technology. Blockchain's capacity for secure and transparent transactions is revolutionizing traditional banking, with applications ranging from cross-border payments and remittances to the issuance of digital currencies by central banks.

9.6.2 SUPPLY CHAIN AND LOGISTICS

Blockchain's transparency and traceability are transforming supply chain and logistics operations. Industries like food, pharmaceuticals, and retail are employing blockchain to track product origins, verify authenticity, and reduce fraud, thereby streamlining the supply chain process.

9.6.3 REAL ESTATE AND PROPERTY MANAGEMENT

Real estate transactions are experiencing a paradigm shift with blockchain. Smart contracts automate property purchases, rental agreements, and property title transfers, dramatically simplifying the process and reducing the need for intermediaries.

9.6.4 HEALTHCARE AND MEDICAL RECORDS

Blockchain technology is enhancing the security and accessibility of healthcare data. Patients have greater control over their medical records, and the technology is facilitating telemedicine and data sharing while ensuring data privacy.

9.6.5 VOTING AND GOVERNANCE

Blockchain-based voting systems are gaining momentum for their ability to ensure secure, transparent, and tamper-resistant elections. Decentralized autonomous organizations (DAOs) are reshaping governance in organizations and communities, creating new avenues for transparency and trust.

9.6.6 ART, MEDIA, AND ENTERTAINMENT

Blockchain's influence is palpable in the creative and entertainment industries. Nonfungible tokens (NFTs) are enabling artists, musicians, writers, and content creators to monetize their digital assets directly, circumventing traditional gatekeepers.

9.6.7 LEGAL AND SMART CONTRACTS

Law firms and legal entities are exploring the automation of legal agreements and dispute resolution through smart contracts. This streamlines legal processes, reduces costs, and minimizes the need for intermediaries.

9.6.8 ENERGY AND SUSTAINABILITY

Blockchain is transforming the energy sector by creating decentralized energy grids. Smart contracts enable energy trading and incentivize renewable energy production, contributing to sustainability and energy efficiency.

9.6.9 EDUCATION AND ACADEMIC RECORDS

Blockchain is revolutionizing the management of academic records, diplomas, and certifications. This technology enhances data security and prevents credential fraud.

9.6.10 GOVERNMENT SERVICES

Governments worldwide are considering blockchain for public services, such as land registry systems, identity management, and secure data sharing. Blockchain's transparent and secure nature can reduce the risk of corruption and fraud in public services.

9.6.11 INSURANCE AND CLAIMS PROCESSING

The insurance industry is turning to blockchain for the automation of claims processing. Smart contracts can trigger compensation when predefined conditions are met, streamlining the process and improving efficiency.

9.6.12 GAMING AND VIRTUAL ASSETS

Blockchain technology is gaining prominence in the gaming industry. Gamers can now own, trade, and sell in-game assets as NFTs, creating new revenue streams for developers and enhancing the gaming experience.

9.6.13 AGRICULTURE AND FOOD SAFETY

Blockchain is playing a pivotal role in improving food safety and traceability in the agriculture industry. Consumers can track the origin of food products, thereby ensuring quality and safety.

9.6.14 CROSS-BORDER TRADE AND E-COMMERCE

Blockchain simplifies cross-border trade and e-commerce by enhancing supply chain management, reducing paperwork, and providing secure and transparent transactions.

9.6.15 INTELLECTUAL PROPERTY

Blockchain technology is employed to protect intellectual property and automate royalty payments, benefiting artists, content creators, and inventors (Wang et al., 2019).

The adoption of ledger technology and smart contracts is catalyzing an industrial transformation, making operations more efficient, reducing costs, and heightening security across a wide array of sectors (Sagayam et al., 2022). As industries increasingly adopt these innovations, the future of ledger development and smart contracts will continue to be interwoven with their expanding influence, opening a landscape teeming with possibilities and opportunities for businesses and individuals alike (Zheng et al., 2020).

9.7   CONCLUSION

In the vast landscape of ledger development and smart contracts, the journey into the future is one marked by boundless potential and transformative innovation. As we conclude this chapter and peer into the horizon, several key takeaways emerge, emphasizing the profound impact these technologies are poised to have on our world.

- **A Path to Decentralization**: The future of ledger development and smart contracts promises a journey towards decentralization. These technologies are unraveling traditional central authorities and intermediaries, offering a transparent, secure, and autonomous path for individuals, businesses, and governments.
- **Industries in Transformation**: The ripple effect of ledger development and smart contract adoption is reshaping industries across the board. From finance to healthcare, supply chain to real estate, the transformative power of blockchain technology is undeniable. It's optimizing operations, streamlining processes, and reducing costs.
- **Inclusive Financial Systems**: The emergence of decentralized finance (DeFi) is reshaping the way we think about traditional banking. Smart contracts are at the core of financial innovations that are making banking services more inclusive, accessible, and cost-effective, especially for those who are underserved by traditional financial institutions.
- **Ownership and Control**: Smart contracts and non-fungible tokens (NFTs) are empowering artists, content creators, and individuals, granting them newfound ownership and control over their digital assets. This shift in the paradigm of ownership is disrupting traditional creative and intellectual property industries.
- **Transparency and Trust**: Trust is a cornerstone of blockchain technology. Through its transparency, immutability, and cryptographic security, it's enabling new levels of trust in various sectors, including voting, governance, and supply chain management.
- **Challenges Ahead**: As with any transformative technology, ledger development and smart contracts are not without their challenges. Scalability,

energy consumption, regulatory hurdles, and security concerns must be addressed for the full realization of their potential.

- **Collaboration and Innovation**: The future of ledger development is a collaborative endeavor. Innovation is thriving through open-source communities, industry partnerships, and interdisciplinary collaboration. The dynamic nature of these technologies fosters creativity and solution-driven thinking.
- **Ethical and Regulatory Considerations**: As the blockchain ecosystem grows, ethical considerations and regulatory frameworks are evolving. Striking the right balance between innovation, privacy, security, and consumer protection remains a top priority.

In this chapter, we have navigated the evolving landscape of ledger development and smart contracts, exploring the technology's potential to reshape industries, enhance transparency, and usher in a new era of decentralization. As we look to the future, we do so with a sense of anticipation, fully aware that this technology is poised to leave an indelible mark on the way we conduct business, govern, and interact in the digital world. While challenges persist, the spirit of innovation and collaboration that defines this space promises a future that is nothing short of extraordinary.

REFERENCES

Barman, A., Barman, M., & Jamader, A. R. (2024). Ecological Impacts of the Himalayan Region in West Bengal. In *Mountain Tourism and Ecological Impacts: Himalayan Region and Beyond* (pp. 47–61). IGI Global.

Das, P., Martin Sagayam, K., Rahaman Jamader, A., & Acharya, B. (2022). Remote Sensing in Public Health Environment: A Review. In *Internet of Things Based Smart Healthcare: Intelligent and Secure Solutions Applying Machine Learning Techniques*, 379–397. https://doi.org/10.1007/978-981-19-1408-9_17

Das, P., Jamader, A. R., Acharya, B. R., & Das, H. (2019, May). HMF based QoS aware recommended resource allocation system in mobile edge computing for IoT. In *2019 International Conference on Intelligent Computing and Control Systems (ICCS)* (pp. 444–449). IEEE.

Hamilton, M. (2020). Blockchain distributed ledger technology: An introduction and focus on smart contracts. *Journal of Corporate Accounting & Finance*, *31*(2), 7–12.

Jamader, A. R., Chowdhary, S., Jha, S. S., & Roy, B. (2023). Application of Economic Models to Green Circumstance for Management of Littoral Area: A Sustainable Tourism Arrangement. *SMART Journal of Business Management Studies*, *19*(1), 70–84.

Jamader, A. R., Das, P., Acharya, B., & Hu, Y. C. (2021). Overview of security and protection techniques for microgrids. In *Microgrids* (pp. 231–253). CRC Press.

Jamader, A. R., Chowdhary, S., & Shankar Jha, S. (2023). A Road Map for Two Decades of Sustainable Tourism Development Framework. In *Resilient and Sustainable Destinations after Disaster: Challenges and Strategies* (pp. 9–18). Emerald Publishing Limited.

Jamader, A. R., Das, P., & Acharya, B. R. (2019, May). BcIoT: Blockchain based DDoS prevention architecture for IoT. In *2019 International Conference on Intelligent Computing and Control Systems (ICCS)* (pp. 377–382). IEEE.

Jamader, A. R. (2022). A Brief Report of the Upcoming & Present Economic Impact To Hospitality Industry In COVID19 Situations. *Journal of Pharmaceutical Negative Results*, 2289–2302. https://doi.org/10.47750/pnr.2022.13.s09.058

Jamader, A. R., Das, P., & Acharya, B. (2022). An analysis of consumers acceptance towards usage of digital payment system, Fintech and CBDC. *Fintech and CBDC (January 1, 2022)*.

Jamader, A. R., Immanuel, J. S., Ebenezer, V., Rakhi, R. A., Sagayam, K. M., & Das, P. (2023). Virtual Education, Training and Internships in Hospitality and Tourism During Covid-19 Situation. *Journal of Pharmaceutical Negative Results*, 286–290. https://doi .org/10.1080/10963758.2021.1907198

Juels, A., Kosba, A., & Shi, E. (2016, October). The ring of gyges: Investigating the future of criminal smart contracts. In *Proceedings of the 2016 ACM SIGSAC Conference on Computer and Communications Security* (pp. 283–295).

Khan, S. N., Loukil, F., Ghedira-Guegan, C., Benkhelifa, E., & Bani-Hani, A. (2021). Blockchain smart contracts: Applications, challenges, and future trends. *Peer-to-peer Networking and Applications*, *14*, 2901–2925.

Kirillov, D., Iakushkin, O., Korkhov, V., & Petrunin, V. (2019). Evaluation of tools for analyzing smart contracts in distributed ledger technologies. In *Computational Science and Its Applications–ICCSA 2019: 19th International Conference, Saint Petersburg, Russia, July 1–4, 2019, Proceedings, Part II 19* (pp. 522–536). Springer International Publishing.

Li, J., & Kassem, M. (2021). Applications of distributed ledger technology (DLT) and Blockchain-enabled smart contracts in construction. *Automation in construction*, *132*, 103955.

Nayak, D. K., Mishra, P., Das, P., Jamader, A. R., & Acharya, B. (2022). Application of deep learning in biomedical informatics and healthcare. In *Smart Healthcare Analytics: State of the Art* (pp. 113–132).

Sagayam, K. M., Das, P., Jamader, A. R., Acharya, B. R., Bonyah, E., & Elngar, A. A. (2022). DeepCOVIDNet: COVID-19 Detection of Chest Image Using Deep Learning Model. https://doi.org/10.21203/rs.3.rs-1725511/v1

Wang, S., Ouyang, L., Yuan, Y., Ni, X., Han, X., & Wang, F. Y. (2019). Blockchain-enabled smart contracts: architecture, applications, and future trends. *IEEE Transactions on Systems, Man, and Cybernetics: Systems*, *49*(11), 2266–2277.

Zheng, Z., Xie, S., Dai, H. N., Chen, W., Chen, X., Weng, J., & Imran, M. (2020). An overview on smart contracts: Challenges, advances and platforms. *Future Generation Computer Systems*, *105*, 475–491.

10 Decoding Cryptocurrency
A Digital Revolution in Transactions

Kyvalya Garikapati,
Akash Bag, and Sambhabi Patnaik

10.1 INTRODUCTION

The legitimacy and authenticity of the currency being traded is a key problem in financial transactions, whether they include actual paper money or digital versions. Recipients must confirm that any traditional paper or fiat currency they receive is legitimate and not a counterfeit. Similarly, as money is transferred electronically from one account to another in digital transactions, banks are crucial in confirming the legitimacy of the involved bank accounts (Reserve Bank of India, 2021). However, some people have imagined a scenario in which banks are no longer a factor in this equation, which is a notion with the potential to reduce the need for intermediaries (Reserve Bank of India, 2021). This motive can be attributed to several things, such as mistrust of banks, excessive transaction costs imposed by financial institutions, or privacy worries brought on by transaction tracking.

Private agreements were developed to ease transactions in response to this desire to avoid the participation of banks and other intermediaries. Electronic money, however, presented a significant problem because it was simple to duplicate and could thus be used again. Double spending is a phenomenon that significantly hampered the feasibility of these alternative transaction mechanisms. Bitcoin was offered as a solution to this problem, and it is based on blockchain technology (Orcutt, 2017). By requiring changes to be made across every consecutive block on the blockchain, transactions made using this technique are broadcast to all computers in the network, presumably eliminating the chance of double spending.

Money is essential for every individual, addressing various human needs through its utilization. Amidst the advent of the Fourth Industrial Revolution marked by modernization and globalization, blockchain technology has emerged, giving rise to cryptocurrency. Cryptocurrency, a decentralized digital currency, signifies a virtual form of money that lacks physical presence, ensuring secure and untraceable transactions.

Varieties of digital currencies, such as Bitcoin, Ethereum, Litecoin, Monero, and others, have evolved from this blockchain development. Despite its intangible nature, cryptocurrency functions similar to traditional currencies, complete with exchange rates that exhibit fluctuations susceptible to exploitation by traders. Transactions in cryptocurrency involve direct online transfers between individuals without the need for intermediaries. While offering efficiency and convenience, cryptocurrency faces challenges, notably the absence of a central authority to address transaction-related issues and the frequent occurrence of money laundering crimes. Effectively leveraging cryptocurrency and blockchain technology in the contemporary era of globalization presents a significant challenge (Amsyar et al., 2020).

The idea of Bitcoin, a decentralized type of virtual currency, is fundamental to this rapidly changing environment. Litecoin, Dogecoin, Ethereum, Binance Coin, and Bitcoin are well-known cryptocurrencies. These virtual currencies run on the foundation of blockchain technology, allowing users to transact privately while enjoying the advantages of cryptographic security measures. From how we handle money to how we trade and reduce risks associated with cryptocurrency use, this innovation has the potential to alter many facets of financial transactions. A distributed ledger system, in which data is arranged into "blocks" and linked together in a chain-like structure, is the basis of blockchain technology. It's important to note that a single entity does not manage this ledger; each user device has access to a private digital database. The infrastructure that supports numerous entities or locations' simultaneous access, validation, and immutable record updates is known as distributed ledger technology (DLT).

The security of these digital transactions is largely achieved by cryptography. Cryptography ensures the secrecy and integrity of data sent between parties and is based on sophisticated mathematical principles. Cryptography includes methods like the Caesar cipher, monoalphabetic cipher, and Playfair cipher that are all intended to prevent unwanted access to or alteration of sent information. There are two main reasons why cryptocurrencies are becoming more widely used and popular (Mehrotra, 2021). First, the simplicity of buying cryptocurrencies has democratized access for everyone wishing to participate in this avant-garde financial environment. Second, the "network effect" (Narayanan et al. 2016) is crucial; the more people use this technology, the more valuable it becomes (Kochkodin, 2017). A popular cryptocurrency site in the US called Coinbase is an illustrative example of this issue. The number of users on Coinbase increased by over 100,000 daily due to the Chicago Mercantile Exchange's announcement regarding introducing Bitcoin futures.

10.2 TYPES OF CRYPTOCURRENCIES

- Payment cryptocurrency.
- Utility tokens.
- Stablecoins.
- Central bank digital currencies (CBDC).

10.2.1 PAYMENT CRYPTOCURRENCY

Payment cryptocurrencies represent a significant category within the realm of digital currencies. Bitcoin, renowned as one of the most well-known cryptocurrencies, stands as the pioneering example of a successful payment cryptocurrency. The primary objective of payment cryptocurrencies, as their name suggests, extends beyond serving as a medium of exchange; they function as exclusively peer-to-peer electronic cash systems designed to facilitate transactions seamlessly. In essence, these cryptocurrencies are tailored to be versatile currencies for various transactions, and as a result, they operate on dedicated blockchains specifically crafted for payment purposes. Notably, these blockchains lack the capability to support smart contracts and decentralized applications (Dapps). Furthermore, payment cryptocurrencies typically have a finite supply of digital coins, contributing to their inherently deflationary nature. The controlled issuance of these digital coins leads to an anticipation of increasing value as the available supply diminishes over time. Exemplary instances of payment cryptocurrencies encompass Bitcoin, Litecoin, Monero, Dogecoin, and Bitcoin Cash.

10.2.2 UTILITY TOKENS

Tokens refer to cryptographic assets that operate on top of an existing blockchain. The Ethereum network pioneered the concept of enabling other cryptographic assets to leverage its blockchain infrastructure. Vitalik Buterin, the creator of Ethereum, envisioned his cryptocurrency as a programmable form of open-source money, designed to facilitate smart contracts and decentralized applications, disrupting traditional financial and legal entities. A notable distinction between tokens and payment cryptocurrencies lies in their supply dynamics. Unlike capped payment cryptocurrencies, tokens such as Ether on the Ethereum network have no maximum limit. Consequently, these cryptocurrencies exhibit inflationary characteristics. The continuous creation of more tokens implies that the value of this digital asset is anticipated to decrease, akin to the currency depreciation observed in countries with a constant influx from their currency printing presses.

10.2.3 STABLECOINS

In response to the volatility observed in numerous digital assets, stablecoins are specifically crafted to serve as a reliable store of value. Their value retention is achieved by their link to one or more fiat currencies, making them exchangeable for these traditional currencies. The most common pegs for stablecoins are the US Dollar or the Euro. The entity overseeing the peg is responsible for holding reserves to ensure the stability of the cryptocurrency. This stability is particularly appealing to investors who may utilize stablecoins as a secure savings instrument or as a means of exchange, allowing for consistent value transfers without the fluctuations seen in other cryptocurrencies. One prominent example of a stablecoin is Tether's USDT, ranking as the third-largest cryptocurrency by market capitalization after Bitcoin and Ether. USDT is pegged to the US Dollar, aiming to maintain a stable value of USD1 each. This stability is maintained through Tether's backing of every USDT with USD1 worth of reserve assets in cash or cash equivalents. However, it's crucial to note that stablecoins operate without government regulation or oversight. This underscores the importance of thorough research and due diligence before investing in stablecoins, including a careful examination of the whitepaper and an understanding of how the stablecoin manages its reserves (DeVries, 2016).

10.2.4 CENTRAL BANK DIGITAL CURRENCIES (CBDC)

Central bank digital currency (CBDC) represents a type of cryptocurrency that is issued by the central banks of various nations. These digital currencies, issued either in token form or through an electronic record linked to the country's domestic currency, are regulated and controlled by the respective central banks. Although the integration of CBDCs into financial systems and monetary policies is still in its nascent stage in many countries, there is potential for wider adoption in the future. Similar to conventional cryptocurrencies, CBDCs utilize blockchain technology, aiming to enhance payment efficiency and potentially reduce transaction costs. While the development of CBDCs is ongoing among central banks globally, some share fundamental principles and technology with established cryptocurrencies like Bitcoin.

The issuance of CBDCs in token form or electronic records for ownership verification aligns them with existing cryptocurrencies. However, CBDCs differ in that they are closely monitored and regulated by the issuing government, relinquishing the advantages of decentralization, pseudonymity, and censorship resistance typically associated with other cryptocurrencies. CBDCs leave a traceable "paper trail" of transactions, enabling governments to enforce taxation and economic rents. On the positive side, in a politically stable and inflation-controlled environment, CBDCs are expected to maintain their value over time, or at least closely track the value of the pegged physical currency. Buyers of CBDCs benefit from the backing of the issuing country's full faith and credit, reducing concerns related to fraud and abuse that have plagued some traditional cryptocurrencies (Fokri, 2021).

10.2.5 FROM DIGITAL LEDGER TO CRYPTOCURRENCY CRAZE

There is a common misperception that the birth of blockchain technology can only be credited to the anonymous author known as "Satoshi Nakamoto," who published the Bitcoin white paper in 2008. However, it is possible to go back much further, all the way to the 1980s, to find the origins of this revolutionary technology. A digital ledger system with a distinctive structure was developed in that era by two pioneers, Stuart Haber and Scott Stornetta. This design ensured that changing just one link in the chain would cause the entire sequence to fall out of place—significant questions developed at the time as personal computing started to permeate the typical American family. The first was how one could have confidence in digital data, which was prone to manipulation. Second, could these changeable digital records be trusted at all in the absence of a centralized authority to control them? Nakamoto added the idea of "mining" to the decentralization theory that Haber and Stornetta had previously developed. People could receive Bitcoin as an incentive by figuring out difficult mathematical riddles connected to transactions within a block. This clever reward system served as a deterrent to tampering. Blockchain technology is safe and resistant to fraud because of the network's interconnection, which means that any attempt to tamper with a single block would cause the entire chain to break.

Although Bitcoin didn't leave the world of computer programming until the late 2000s, this technology went in intriguing new directions. The tale of Florida-based computer programmer Laszlo Hanyecz serves as a prime illustration. A remarkable 10,000 Bitcoins were offered by Hanyecz in 2010 in exchange for the simple pleasure of receiving two pizzas delivered to his home (Jones, 2022).The pizzas then went for a reasonable $30 each. But by May 2018, those identical pizzas, sent to Hanyecz by a person in England, had accumulated an incredible worth of USD82 million (Whitaker, 2019). This crucial event took place on May 22, and that day is now recognized as "Bitcoin Pizza Day" (Whitaker, 2019). The laws of supply and demand govern the value of Bitcoin, just like they do for traditional commodities. It's important to understand the path that cryptocurrencies have taken, from their conception in specialized computer programming circles to their incorporation into the lives of regular people. Comprehending this trip becomes essential when considering the potential uses of cryptocurrencies shortly. However, as with every game-changing development, it's critical to recognize and handle the various worries and difficulties that emerge alongside this technological advancement. As time goes on, realizing cryptocurrencies' full potential necessitates striking a fine balance between the concerns they raise and the promises they provide.

10.3   DECRYPTING CRYPTOCURRENCY'S REGULATORY CONUNDRUM

A puzzling conundrum remains in the complex world of cryptocurrencies: why do governments and central banks hesitate to accept this progressive financial system? The main problem is that a formal regulatory framework does not govern cryptocurrency transactions. The decentralized structure of Bitcoin transactions creates concerns due to the lack of channels for addressing issues, in contrast to traditional financial systems where middlemen assure accountability. Let's explore the regulatory worries that have sparked this reluctance to explain this situation. The lack of intrinsic value in cryptocurrencies is one of the main worries.

Contrary to real commodities like gold, the value of cryptocurrencies is purely based on the acceptance and trust given to them by users. Advocates contend that the restricted supply of some cryptocurrencies mitigates this problem due to their controlled production or "mining." However, the variety of new 'cryptocurrencies' threatens to undermine such restrictions. The murky legal position of cryptocurrencies in some international jurisdictions is a further urgent concern. The classification of these digital objects as securities, commodities, currencies, or assets is still up for debate. Although several nations, such as Japan and Switzerland, have enacted legislation encouraging the use of cryptocurrencies, India is torn between outright banning them and incorporating them into its financial system.

When we examine the Indian environment, we encounter fascinating discussions about how to classify cryptocurrencies legally. The tricky balancing act between definitions is essential to treating cryptocurrencies as securities. Cryptocurrencies are not specifically considered securities under the Securities Contract Regulation Act, 1956 (the "SCRA"). However, the more inclusive term "other marketable security" may apply, covering them. Nonetheless, the requirement that a corporate body issue these securities runs against the fundamental anonymity of cryptocurrencies and thereby exempts them from the SCRA's jurisdiction (The Reserve Bank of India Act, 1934).

We are drawn to the definition of "currency" in section 2(h) of the Foreign Exchange Management Act, 1999 (the "FEMA") when we consider cryptocurrencies to be virtual money (Baxter, 2016). The lack of an issuer offering a similar value makes it difficult to align with conventional fiat currency, and classifying cryptocurrencies as "goods" or commodities also has drawbacks. Money and securities are not considered "goods" under the Securities Contract Regulation Act, and it is difficult to categorize cryptocurrencies with certainty due to the ambiguity surrounding them (Narayanan et al., 2016). The unmatched privacy and decentralization of cryptocurrencies are what make them so alluring. Transaction privacy is unmatched by any other identity management system. One's actions within this system eventually result in a succession of recognizable patterns. Therefore, a word of caution does emerge. This progression of behaviors that unintentionally betray the user's real-world identity may compromise the original perceived anonymity.

A fundamental concern about cryptocurrencies' inherent resistance to government regulation emerges amidst these difficulties. Adopting more stable currencies like the US Dollar or the Euro could result from "Dollarization" if cryptocurrencies become a viable alternative to fiat money. Due to transactional anonymity, this change jeopardizes the effectiveness of the central bank's monetary policy and facilitates illegal activity. As a result, controlling the money supply and interest rates to fight inflation becomes difficult, making the present legal structure unsuitable for handling these problems (Narayanan et al., 2016).

10.4   CRYPTOCURRENCY REGULATIONS UNVEILED: A EUROPEAN PERSPECTIVE

The story of cryptocurrency's voyage through the European Union (EU) is one of caution, adaptation, and changing viewpoints. To understand this story better, let's look at the trends influencing Bitcoin rules in the area. The European Banking Authority (EBA) warned about virtual currencies like Bitcoin around 2013. They expressed concern about the risks of purchasing, holding, or trading these digital assets. These dangers included the possibility of financial loss or theft to users, using virtual money for illegal purposes, and the erratic nature of such forms of digital wealth. The EBA took things further in July 2014 by detailing more than 70 potential risks of managing cryptocurrencies across multiple categories. The use of cryptocurrency in financial crimes was, in particular, one of the most urgent concerns. In response to these worries, the EBA provided suggestions aimed at reducing these risks, such as:

1. Encourage national regulatory agencies to discourage financial institutions from transacting in virtual currency.
2. Recommending the designation of organizations working at the intersection of real-world and digital currencies, like virtual currency exchanges, as "obligated entities." The EU Anti-Money Laundering Directive's anti-money laundering and counterterrorist financing rules would apply to them due to this designation.

Germany started a process of integrating cryptocurrencies into its domestic regulatory system. This process included formally designating crypto-assets as financial instruments and licensing Bitcoin custody service providers. Additionally, private funds now have the option to invest up to 20% of their funds in cryptocurrency (Reuters, 2013). In 2020, the United Kingdom (UK) declared that cryptocurrency assets should be considered property (Kahl, 2021). However, it's significant that the UK does not have any cryptocurrency legislation or grant cryptocurrencies the status of legal cash. In March 2022, the Financial Conduct Authority (FCA) of the UK made a startling announcement: all cryptocurrency ATMs in the nation were considered illegal and demanded closure. This position has changed over time (Comply Advantage, 2018b). The 81 similar ATMs that Coin ATM Radar had at the time were affected by this action. Know your customer (KYC) laws, which are intended to prevent money laundering, were not followed, according to the FCA. The FCA also voiced concern about consumer risks due to a lack of regulatory monitoring and security.

10.5   EU REGULATION ON CRYPTO-ASSETS AND THE PROSPECT OF A DIGITAL EURO

The European Union (EU) has made great strides toward establishing the legal framework for cryptocurrency assets and investigating the possibility of a virtual euro (Comply Advantage, 2018b). The Markets in Crypto-Assets Regulation (MiCA), a new piece of legislation, is of utmost significance in this evolving field.

10.5.1   KEY PROVISIONS OF THE MiCA

The EU's strategy for regulating cryptocurrency assets has reached a turning point with the Markets in Crypto Assets Regulation (MiCA). The European Commission unveiled this comprehensive framework in September 2020 as a larger digital finance plan component. MiCA stands out since it directly applies to every EU member state, eliminating the need for separate national implementation laws. By fostering consumer safety and ensuring streamlined access to the developing crypto-asset markets within the United European market, this strategic alignment aims to benefit everyone in the region (Schickler, 2022).

Importantly, MiCA accomplishes the following four basic goals:

1. Creating Legal Clarity: MiCA establishes a solid legal framework for crypto-assets not previously covered by existing financial services legislation, filling a regulatory gap.
2. Fostering Creativity and Fair Competition: By creating a framework that balances safety and creativity, the law hopes to encourage the growth of crypto-assets.
3. Protecting investors, consumers, and market integrity: MiCA aims to shield multiple stakeholders from potential harm while acknowledging the inherent risks associated with crypto-assets.
4. Securing Financial Stability: MiCA includes protections to reduce any potential risks to the security of financial systems arising from the changing cryptographic landscape (Boucheta & Joseph, 2023).

10.5.2   DEVELOPING THE DIGITAL EURO

The idea of a digital euro has also garnered popularity as a contemporary update to traditional money. The Eurosystem, which consists of the European Central Bank (ECB) and the national central banks of eurozone countries, would issue this digital form of money, which would resemble actual euro banknotes. This proposal, which aims to be accessible to citizens and enterprises, is being investigated under the auspices of the digital euro project, which began its investigation phase in October 2021. The objective of this phase, which will last for 24 months until October 2023, is to examine significant issues related to the creation and distribution of the digital euro. A realization phase, in which the ECB's Governing Council would create and test the technological and operational features of the digital euro, might begin after the investigation's conclusion (Boucheta & Joseph, 2023). The high-level task force on central bank digital currency (HLTF-CBDC), which comprises officials from the ECB and national central banks in the eurozone, leads these initiatives (Smith-Meyer, 2022).

10.5.3   IMPLICATIONS AND THINGS TO THINK ABOUT

The proposed legislation and the digital euro proposal have raised enthusiasm and trepidation. Supporters see MiCA as crucial in promoting responsible growth in the Bitcoin business and improving regulatory clarity. In the meantime, launching a digital euro raises questions about its effects during financial unrest, since people might quickly change their deposits into the digital currency supported by the central bank, potentially escalating online bank runs.

10.5.4   UNDERSTANDING CRYPTOCURRENCY REGULATIONS IN THE UNITED STATES

Compared to several other countries, the United States takes a more liberal approach to Bitcoin. In the US, cryptocurrency is recognized as legal because it is covered by the Bank Secrecy Act (BSA). Companies that offer Bitcoin exchange services must join the Financial Crimes Enforcement Network (FinCEN) and set up a strong Anti-Money Laundering and Counter-Terrorism Financing (AML/CFT) program. This guarantees that illegal financial activity is prevented. On March 9, 2022, President Joseph Biden signed an executive order directing the federal government to develop plans for regulating cryptocurrencies, marking a crucial turning point (The White House, 2022a). The Biden administration previously focused primarily on stablecoins to address the possible concerns associated with their value inflation (Murry, 2022). To contextualize this executive order, it is crucial to consider the bipartisan infrastructure legislation of 2021, which outlines a comprehensive national strategy for digital assets. There are six essential components to this strategy (The White House, 2022b):

1. Protecting investment and customers.
2. Ensuring a stable economy.
3. Reducing illegal financial activity.
4. Maintaining US dominance in the world of finance and enhancing economic competitiveness.
5. Encouraging monetary inclusion.
6. We are promoting ethical innovation.

This new set of regulatory measures is anticipated to give tax authorities and banking regulators better tools for policing Bitcoin use (Nakajima, 2022). This regulation also forces cryptocurrency exchanges to keep thorough records of gains and losses from digital asset transactions. The Financial Crimes Enforcement Network (FinCEN) is one example of how the US continues to be proactive in its efforts to write federal cryptocurrency legislation (Financial Crimes Enforcement Network, 2013). FinCEN recognizes cryptocurrency exchanges as money transmitters even though it does not classify cryptocurrencies as legal tender, attributing this position to the idea that cryptocurrency tokens act as "other value that substitutes for currency." Similarly, the Internal Revenue Service (IRS) describes cryptocurrencies as a digital expression of value that serves as a medium of exchange, a unit of account, and a store of value but does not recognize it as legal tender. As a result, the IRS has released tax regulations corresponding to this description (U.S. Government Accountability Office, 2020).

Due to the US Securities and Exchange Commission's (SEC) interpretation that cryptocurrencies are securities, digital wallets and exchanges are now subject to stringent securities laws. Contrarily, because Bitcoin is recognized by the Commodities Futures Trading Commission (CFTC) as a "commodity," trading in cryptocurrency derivatives is permitted on open exchanges. In a similar spirit, FinCEN has clarified that it expects Bitcoin exchanges to abide by the "Travel Rule."

This involves gathering and disseminating data about the initiators and beneficiaries of cryptocurrency transactions. The Bank Secrecy Act's regulatory framework, which includes its interpretation of the Travel Rule, is applied to these exchanges similarly to conventional money transmitters. A crucial turning point in this regulatory path occurred in October 2020 with the publication by FinCEN of a Notice of Proposed Rulemaking (NPRM) regarding changes to the Travel Rule. With this, new regulatory requirements for Bitcoin exchanges were introduced (Comply Advantage, 2018a). As the US works through the complexity of cryptocurrency legislation, the regulatory environment is poised to change how digital assets are used in the country's financial system in the future.

10.6 CHINA'S CRYPTOCURRENCY BAN: WHY IT HAPPENED AND ITS IMPACT

The People's Bank of China (PBOC) took a huge step back in September 2021 by outlawing all cryptocurrency transactions. This choice was made as a result of a few major issues. First, the PBOC raised concerns about the security and legality of cryptocurrencies by pointing out that they were being used as a tool for financial crimes. The PBOC also emphasized the significant risks associated with the speculative nature of cryptocurrencies, which could endanger China's financial stability. The endeavor to stop capital flight from China may also have played a role. Over USD50 billion in cryptocurrencies had reportedly fled East Asian accounts between 2019 and 2020, according to Chainalysis, an analytics firm specializing in cryptocurrency movements, possibly due to capital flight.

Several governments have initially adopted a non-interventionist stance toward cryptocurrencies, yet the swift growth and transformation of the crypto landscape, along with the emergence of decentralized finance (DeFi), have compelled regulators to formulate guidelines for this burgeoning sector. Regulatory frameworks exhibit significant diversity globally, with some governments endorsing cryptocurrencies while others outright prohibit them. The task facing regulators, according to experts, is to establish rules that mitigate conventional financial risks without impeding the progress of innovation.

The Basel Committee on Banking Supervision has made substantial progress in global cryptocurrency cooperation. A framework for regulating crypto-assets by January 1, 2025, is laid forth in an article they published titled "Prudential treatment of crypto-asset exposures" (Basel Committee on Banking Supervision, 2022) This study divides crypto-assets into two main categories, Group 1 and Group 2. Each group is broken into two smaller groups, with distinct standards defining each category. Traditional assets that have been tokenized, such as digital bonds or loans, are included in Group 1 crypto-assets (Group 1a). Group 1b, which consists of crypto-assets with procedures to stabilize their value and make them redeemable for a predefined quantity of reference assets, is another category. These assets imitate the risk management qualities of traditional assets. There is a test for these Group 1b crypto-assets to ensure they satisfy the requirements (Basel Committee on Banking Supervision, 2022).

Based on the current Basel Framework, the regulatory framework specifies capital requirements for Group 1 crypto-assets based on the risk involved with the underlying assets (Basel Committee on Banking Supervision, 2022). Group 2 crypto-assets, on the other hand, don't fit the criteria outlined before. They are typically unbacked and present greater risks than Group 1 crypto-assets. They receive more cautious capital handling as a result (Basel Committee on Banking Supervision, 2022). There are two additional categories within Group 2: Group 2a and Group 2b. Group 2a consists of crypto-assets that allow for some degree of hedging. Only Group 1a crypto-assets are designated High-Quality Liquid Assets, indicating their high stability and potential value. The article shows that this framework will not cover central bank digital currencies. The research also details capital and liquidity requirements based on the credit risk and market risk associated with these crypto-assets, depending on their categorization and limits on bank exposure to Group 2 assets.

10.7   PARTICIPATING ON A GLOBAL SCALE

India has recognized these suggested measures as a member of this committee, demonstrating its engagement in establishing future international legislation for cryptocurrencies. This proactive involvement represents the growing worldwide impetus to govern the complex world of cryptocurrencies logically and securely.

10.8   THE RBI'S STANCE ON CRYPTOCURRENCIES

Cryptocurrencies and other digital assets, often known as virtual currencies, have raised concerns for the Reserve Bank of India (RBI). The RBI first discussed these virtual currencies and outlined the hazards connected with their use in a press release in 2013 (Reserve Bank of India, 2016). In the RBI's Financial Stability Report from December 2016, the bank raised concerns about data security, consumer protection, and potential effects on monetary policy (Money Control, 2023).

The RBI issued another warning in February 2017, urging people not to use virtual currencies. Following this caution, the Ministry of Finance established an interdepartmental committee in April 2017 that concluded that virtual currencies should not be used as legal cash (Reserve Bank of India, 2016). The research downplayed the notion of virtual currencies serving as money but recognized the importance of blockchain technology. Building on earlier statements, the RBI published a circular in April 2018 indicating that organizations it supervised were no longer allowed to deal in virtual currencies or provide services for transactions using them—the action aimed to reduce the alleged hazards related to these currencies (PRS, 2019).

However, this Circular was subject to legal review (Reserve Bank of India, 2018). In a dispute against the RBI, the Internet and Mobile Association of India contested its constitutionality. The RBI was accused of exceeding its jurisdiction by outlawing virtual currencies, which was thought to violate Article 19(1)(g) of the Indian Constitution's guarantee of the right to trade and conduct business. The RBI said that the prohibition was constitutionally permissible and that virtual currencies fell under its authority (Bhardwaj, 2020). The RBI also emphasized its worries about responsibility and the potential abuse of virtual currencies for illicit purposes. Ultimately, the Supreme Court determined that the April 2018 Circular breached the Doctrine of Proportionality and violated the rights guaranteed by Article 19(1)(g). In other words, the court made the case that even while virtual currencies may not be officially recognized as legal currency, they can nonetheless act as real money and, as a result, deserve consideration and protection.

The Ministry of Corporate Affairs responded to this decision by issuing a notification in March 2021 requiring businesses to declare their virtual currency transactions (James, 2022). The RBI then revoked its April 2018 Circular in May 2021 and instructed cryptocurrency trading platforms to abide by Foreign Exchange Management Act (FEMA), know your customer (KYC), and anti-money laundering standards. As it navigates the complexity of a digital financial ecosystem and attempts to balance innovation, security, and legality, the RBI is developing its stance on cryptocurrencies, highlighted by this journey of caution and regulation (Singh, 2021).

10.9 NAVIGATING CRYPTOCURRENCIES AND DIGITAL RUPEE: A TALE OF TAXATION, CBDCS, AND POTENTIAL DISRUPTION

The phrase "digital assets" has recently gained popularity and now refers to various digital holdings other than cryptocurrency. In addition to regulating cryptocurrencies and non-fungible tokens, this extended definition gives governments the authority to control new digital assets. The Indian government has introduced measures to manage taxation in this area due to the cryptocurrency industry's changing landscape. Adding Section 115BBH to the Income Tax Act, which levies a 30% tax (minus surcharge and cess) on income from transfers of virtual digital assets, is a crucial step.[1],[2] A 1% tax deduction at source was implemented for Bitcoin transactions beginning July 1, 2022, to simplify taxation.[3] However, this legislative move has raised questions since analysts believe it may harm India's developing cryptocurrency market and cause a daily transaction volume reduction of more than 70%.[4]

On November 29, 2022, the Reserve Bank of India (RBI) presented its innovative pilot project for a retail digital rupee (Das, 2022). This plan establishes digital rupee denominations corresponding to paper money to be used for transactions through digital wallets provided by partner banks and is set to go into effect on December 1, 2022.[5] Utilizing the effectiveness and speed of digital transactions, the central bank digital currency (CBDC) project of the RBI, often known as the digital rupee, seeks to supplement the current range of payment methods. The two main goals of the RBI's CBDC strategy are to closely resemble the features of paper money and provide a seamless integration of the rupee into the digital economy. With a progressive deployment plan, the RBI is looking into an account-based CBDC for the wholesale industry and a token-based CBDC for the retail segment. However, this project demands addressing the financial industry's regulatory evasion problems related to cryptocurrency.

The article highlights the transformational potential of cryptocurrencies within the financial sector as it considers the future. For areas with restricted access to banking services, cryptocurrencies provide a peer-to-peer system that gets around traditional banking restrictions (DeVries, 2016). In contrast to the drawn-out delays of traditional banking procedures, these digital currencies enable quick cross-border transactions. The advantage of cryptocurrencies lies in the lower transaction fees and the increased efficiency of digital transactions (Williams, 2021). Cryptocurrencies allow for fast transactions since they are not constrained by the checks and balances built into banking systems (Wątorek et al., 2023). In addition, the technology's quickness and effectiveness encourage creative applications like trading cyber-assets, with recent examples like Gucci's sale of digital replicas of its products illustrating the market's interest.

The article compares the possible influence of cryptocurrency on energy usage to the revolutionary effects of the internet on data interchange to address environmental concerns. Cryptocurrencies could be used to offset the existing energy-intensive mining procedures and the infrastructure requirements of traditional banking, in line with sustainable energy goals. Additionally, democratizing crowdfunding and enhancing financial inclusion are two other promising uses of cryptocurrencies that could support social and economic growth on a global scale. The author emphasizes that it is limiting to only see cryptocurrencies as money, though. Cryptocurrencies can enable real-time transaction tracking and do away with the necessity for escrow accounts due to their dual nature of transparency and capitalism (Finlay, 2021). However, difficulties still exist because governmental reticence and the speculative nature of cryptocurrency investing fuel market volatility, which might destabilize economies (Catlow et al., 2017).

The risk of such adoption is shown by the example of El Salvador, the first nation to accept Bitcoin as legal cash. Bitcoin's value fell amidst significant investment, impacting the economy. The Salvadoran example is a cautionary tale, highlighting the necessity for careful deliberation as countries like India consider adopting cryptocurrencies (Catlow et al., 2017).

10.10   CONCLUSION

International cooperation is needed to create effective rules for cryptocurrencies, those transnational digital assets. The Reserve Bank of India's (RBI) December 2022 Financial Stability Report emphasized the influence of cryptocurrencies on a global scale (Acosta, 2022). Due to recent market volatility and the rising integration of cryptocurrencies with traditional finance, immediate action is required. The Financial Stability Board, which supports global consistency in cryptocurrency regulation based on the tenet of "same activity, same risk, same regulation," is a significant participant in this discussion. The study also emphasizes the Basel Committee's division of cryptocurrency assets into two categories and the necessity of bolstering financial stability while preparing for issues that may come up during India's G-20 Presidency (PRS, 2021). This alliance of 20 nations seeks to create uniform cryptocurrency legislation at the international level.

Prospects for cryptocurrencies depend on how they will affect society and politics. The world is investigating how these technological developments may alter our daily lives. This change has the potential to be advantageous for both people and nations. However, development requires an amicable agreement between all parties, including business owners, IT companies, customers, investors, and financial institutions. Without a defined regulatory framework, reaching this consensus is difficult (PRS, 2021). With cryptocurrencies, a new era of ownership freedom has begun. Although they maintain data integrity and permit quick transactions without intermediaries, their usefulness is still elusive.

In contrast to conventional currencies controlled by central banks, cryptocurrencies derive their value from user consensus. Despite being liberating, this decentralization may also present risks given the wide range of transactions made possible by cryptocurrencies. Regulation is the answer; a complete ban or disdain is unfeasible (The Hindu Bureau, 2022). The proponents of cryptocurrencies and the central banks trying to regulate them fight. Central banks must acknowledge the persistent presence of cryptocurrencies.

Similarly, proponents of autonomous financial systems should admit that central bank participation boosts credibility and reduces volatility. In this case, it could be wise for central banks to cautiously legitimize cryptocurrencies, which would benefit all parties. Unprecedented issues that go beyond traditional regulations are undoubtedly introduced by the incredible flexibility that this technology offers. It would be naive to discount the benefits of new technology, though. With all its complications, accepting new technology is a journey that cannot be stopped.

NOTES

1. https://www.hodlbot.io/blog/types-of-cryptocurrencies
2. Sec. 2(47A), Income Tax Act, 1961.
3. Sec. 115BBH, Income Tax Act, 1961.
4. Sec. 194S, Income Tax Act, 1961.
5. Ibid.

REFERENCES

Acosta, S. (2022, September 11). One year on, El Salvador's Bitcoin experiment has proven a spectacular failure. *The Conversation*. http://theconversation.com/one-year-on-el-salvadors-bitcoin-experiment-has-proven-a-spectacular-failure-190229

Amsyar, I., Christopher, E., Dithi, A., Khan, A., & Maulana, S. (2020). The Challenge of Cryptocurrency in the Era of the Digital Revolution: A Review of Systematic Literature. *Aptisi Transactions on Technopreneurship (ATT)*, 2, 153–159. https://doi.org/10.34306/att.v2i2.96

Basel Committee on Banking Supervision. (2022). *Prudential treatment of cryptoasset exposures.* https://www.bis.org/bcbs/publ/d545.htm

Baxter, L. (2016). Adaptive Financial Regulation and RegTech: A Concept Article on Realistic Protection for Victims of Bank Failures. *Duke Law Journal*, *66*(3), 567–604.

Bhardwaj, P. (2020, March 4). SC quashes RBI's ban on Cryptocurrency trading [Full Report]. *SCC Times.* https://www.scconline.com/blog/post/2020/03/04/sc-quashes-rbis-ban-on -cryptocurrency-trading/

Boucheta, H., & Joseph, A. (2023, April 26). *MiCA – Markets in Crypto-Assets Regulation Memo.* Securities Services. https://securities.cib.bnpparibas/markets-in-crypto-assets -regulation/

Buchholz, K. (2022, March 18). *Infographic: Where the World Regulates Cryptocurrency.* Statista Daily Data. https://www.statista.com/chart/27069/cryptocurrency-regulation -world-map

Catlow, R., Garrett, M., Jones, N., & Skinner, S. (2017). *Artists Re:thinking the Blockchain.* Torque Editions.

Comply Advantage. (2018a, July 5). *Cryptocurrency Regulations Around the World.* ComplyAdvantage. https://complyadvantage.com/insights/cryptocurrency-regulations -around-world/

Comply Advantage. (2018b, July 6). *Crypto Regulations in the UK.* ComplyAdvantage. https://complyadvantage.com/insights/cryptocurrency-regulations-around-world/cryp-tocurrency-regulations-uk-united-kingdom/

Das, S. (2022, November 3). *High tax on crypto may kill industry in India: Binance CEO.* Mint. https://www.livemint.com/market/cryptocurrency/high-tax-on-crypto-may-kill -industry-in-india-binance-ceo-11667478028911.html

DeVries, P. (2016). An Analysis of Cryptocurrency, Bitcoin, and the Future. *International Journal of Business Management and Commerce*, *1*, 1–9.

Dolan, S. (2020, March 3). *How the Laws & Regulations Affecting Blockchain Technology and Cryptocurrencies, Like Bitcoin, Can Impact Its Adoption | Business Insider India.* Business Insider. https://www.businessinsider.in/finance/news/how-the-laws-regula-tions-affecting-blockchain-technology-and-cryptocurrencies-like-bitcoin-can-impact -its-adoption/articleshow/74464680.cms

Financial Crimes Enforcement Network. (2013, March 18). *Application of FinCEN's Regulations to Persons Administering, Exchanging, or Using Virtual Currencies.* FinCEN.Gov. https://www.fincen.gov/resources/statutes-regulations/guidance/applica-tion-fincens-regulations-persons-administering

Finlay, R. (2021, December 11). *5 Ways Cryptocurrency Will Change the World of Commercial Real Estate.* Entrepreneur. https://www.entrepreneur.com/money-finance/5-ways-cryp-tocurrency-will-change-the-world-of-commercial/397259

Fokri, E. (2021). Classification of Cryptocurrency: A Review of the Literature. *Turkish Journal of Computer and Mathematics Education (TURCOMAT)*, *12*, 1353–1360. https://doi.org/10.17762/turcomat.v12i5.2027

James, J. (2022, March 19). *Changes in Financial Reporting Disclosures of Companies.* James & James. https://www.jjglobal.in/insight/changes-in-financial-reporting--disclo-sures-of-companies

Jones, E. (2022, September 21). *A Brief History of Cryptocurrency.* CryptoVantage. https:// www.cryptovantage.com/guides/a-brief-history-of-cryptocurrency/

Kahl, S. (2021, July 30). *Germany to Allow Institutional Funds to Hold up to 20% in Crypto.* Bloomberg.Com. https://www.bloomberg.com/news/articles/2021-07-30/germany-to -allow-institutional-funds-to-hold-up-to-20-in-crypto

Kelleher, J. p. (2023, October 8). *Why Do Bitcoins Have Value?* Investopedia. https://www .investopedia.com/ask/answers/100314/why-do-bitcoins-have-value.asp

Kochkodin, B. (2017, November 2). *One of the Biggest Bitcoin Exchanges Just Added 100,000 Users in a Single Day.* Bloomberg.Com. https://www.bloomberg.com/news/articles /2017-11-02/bitcoin-exchange-added-100-000-users-in-a-day-as-price-exploded

Mehrotra, U. (2021, November 3). Cryptocurrency: A Regulatory Conundrum. *SCC Times.* https://www.scconline.com/blog/post/2021/11/03/cryptocurrency-a-regulatory -conundrum/

Money Control. (2023, December 28). *RBI Financial Stability Report Shows Indian Banks Have Improved Health, Capital Prepardeness.* Moneycontrol. https://www.moneycon-trol.com/news/business/rbi-financial-stability-report-banks-gross-npas-falls-to-3-2-in -sept-net-npas-down-to-0-8-11968191.html

Murry, C. (2022, May 20). *Cryptocurrency Explained With Pros and Cons for Investment.* Investopedia. https://www.investopedia.com/terms/c/cryptocurrency.asp

Nakajima, D. (2022, March 28). *Cryptocurrency: Regulations in the USA on the way.* ECOVIS International. https://www.ecovis.com/global/cryptocurrency-regulations-in-usa/

Narayanan, A., Bonneau, J., Felten, E., Miller, A., & Goldfeder, S. (2016). *Bitcoin and Cryptocurrency Technologies: A Comprehensive Introduction.* Princeton University Press.

Orcutt, M. (2017, December 14). *A Cryptocurrency Without a Blockchain Has Been Built to Outperform Bitcoin.* MIT Technology Review. https://www.technologyreview.com /2017/12/14/104996/a-cryptocurrency-without-a-blockchain-has-been-built-to-outper-form-bitcoin/

PRS. (2019, February 28). *Committee Reports.* PRS Legislative Research. https://prsindia.org /policy/report-summaries/virtual-currencies-india

PRS. (2021). *State of State Finances: 2023–24.* PRS Legislative Research. https://prsindia .org/budgets/states/policy/state-of-state-finances-2023-24

Reserve Bank of India. (2016, July 19). *RBI cautions users of Virtual Currencies against Risks.* https://press.princeton.edu/books/hardcover/9780691171692/bitcoin-and-cryp-tocurrency-technologies

Reserve Bank of India. (2018, June 4). *Prohibition on dealing in Virtual Currencies (VCs).* https://www.rbi.org.in/Commonperson/english/scripts/Notification.aspx?Id=2632

Reserve Bank of India. (2021). *Responsible Digital Innovation—T. Rabi Sankar.* https://www .rbi.org.in/Scripts/BS_ViewBulletin.aspx?Id=20564

Reuters. (2013, December 13). *EU Banking Watchdog Warns of Risks from Bitcoin.* Reuters. https://www.reuters.com/article/idUSL6N0JR35V/

Schickler, J. (2022, October 5). *EU Seals Text of Landmark Crypto Law MiCA, Fund Transfer Rules.* https://www.coindesk.com/policy/2022/10/05/eu-seals-text-of-landmark-crypto -law-mica-fund-transfer-rules/

Singh, T. (2021, June 1). RBI issues Customer Due Diligence for transactions in Virtual Currencies (VC). *SCC Times.* https://www.scconline.com/blog/post/2021/06/01/rbi -issues-customer-due-diligence-for-transactions-in-virtual-currencies-vc/

Smith-Meyer, B. (2022, February 9). *Digital euro bill due early 2023.* POLITICO. https:// www.politico.eu/article/digital-euro-bill-due-early-2023/

The Hindu Bureau. (2022, December 11). India to Seek Harmony in Crypto Regulation in G-20 Finance Talks. *The Hindu.* https://www.thehindu.com/business/Economy/india -to-seek-harmony-in-crypto-regulation-in-g-20-finance-talks/article66252067.ece

The Reserve Bank of India Act, Pub. L. No. (Act, No. 02 of 1934), S. 2(aiv) (1934).

The White House. (2022a, March 9). *Executive Order on Ensuring Responsible Development of Digital Assets.* The White House. https://www.whitehouse.gov/briefing-room/presi-dential-actions/2022/03/09/executive-order-on-ensuring-responsible-development-of -digital-assets/

The White House. (2022b, November 15). *FACT SHEET: One Year into Implementation of Bipartisan Infrastructure Law, Biden-Harris Administration Celebrates Major Progress in Building a Better America*. The White House. https://www.whitehouse.gov/briefing-room/statements-releases/2022/11/15/fact-sheet-one-year-into-implementation-of-bipartisan-infrastructure-law-biden-harris-administration-celebrates-major-progress-in-building-a-better-america/

U. S. Government Accountability Office. (2020, March 10). *Virtual Currencies: Additional Information Reporting and Clarified Guidance Could Improve Tax Compliance*. U.S. GAO. https://www.gao.gov/products/gao-20-188

Wątorek, M., Kwapień, J., & Drożdż, S. (2023). Cryptocurrencies Are Becoming Part of the World Global Financial Market. *Entropy*, *25*(2), 377. https://doi.org/10.3390/e25020377

Whitaker, A. (2019). Art and Blockchain: A Primer, History, and Taxonomy of Blockchain Use Cases in the Arts. *Artivate: A Journal of Enterprise in the Arts*, *8*, 21–47. https://doi.org/10.34053/artivate.8.2.2

Williams, G. A. (2021, May 31). *Will Gucci's Digital Bag Disrupt Luxury?* Jing Daily. https://jingdaily.com/posts/gucci-roblox-dionysus-digital-fashion

11 The Advantage of Security Using Blockchain Technology for Medical Tourism
Approaches towards Technologies for Healthcare 4.0

Santanu Dasgupta, Biswaranjan Acharya, and Asik Rahaman Jamader

11.1 INTRODUCTION

Medical tourism is a fast-growing travel concept that has recently achieved immense appeal. Every year, an increasing number of people travel overseas to acquire medical care. Despite the difficulties in confirming accurate numbers for medical tourism, the global number of medical tourists in 2019 was projected to be 23,043,000, with a projected increase to 70,359,000 by 2027. The worldwide medical tourism sector was valued at around US$104.68 billion in 2019 and is expected to reach US$273.72 billion by 2027. Based on a research released, due to the COVID-19 pandemic, these forecasts are likely to be revised downward. Affordable healthcare technologies are one of the main elements driving the recent rise of the medical tourism business. Conveyance at a cheaper price, in addition to increasing advertising of countries as destinations for health tourism, just because this huge technology provides a valuable economic opportunity for such places, the number of countries presenting themselves as medical tourism destinations has increased dramatically (Barkan, M., & Tapliashvili, N., 2018).

The significance of data technology in the tourist business cannot be overstated. With the rising digitalization of healthcare, medical tourism destinations and medical service providers may employ technological advancements to create individualized healthcare delivery services and encourage new visitors. The internet has

DOI: 10.1201/9781003453109-11

been shown to play a vital role in giving information to prospective medical tourists and linking them with hospitals and doctors. Previous research has shown that advanced technologies may be economically incorporated into medical tourism, including travel agencies' process of building and sharing, electronic health data platforms, or stage structures for aesthetic procedures. Blockchain technology is regarded as a fundamental innovation because it has the ability to underpin new financial and social institutions. Blockchain technology and cryptocurrencies have recently emerged as hot topics in tourism and tourism-related studies. Furthermore, in terms of medical tourism research, there are just a few available publications that examine the effects of distributed ledger technology on healthcare services. These publications address several unresolved problems in health tourism involving trusting relationships, treatment and risk transparency, and medical record privacy by concentrating on important blockchain aspects. Blockchain technology, for example, can help medical tourism flourish indefinitely (Das et al., 2019).

Blockchain technology may help promote decentralized travel solutions and better administration of electronic health record systems, as well as help construct smarter medical gadgets and assist in the issuance of secure protocol healthcare certifications. Ref investigates the potential for blockchain to safeguard patient information located in a web infrastructure. According to Ref, blockchain may assure fragmentation in the medical tourism business, promote clarity in communication, trust, and transparency, ameliorate a significant problem of escalating costs, and facilitate stakeholder engagement. Ref emphasizes the significance of Bitcoin use in the healthcare tourism sector as well as quantitatively establishes a favourable association among cryptocurrency use and wellness tourism ambitions. However, additional scientific study is required to have a complete understanding of the effects of blockchain technology on international care. The health sector is divided into different stages: before procedure and after procedure (Das et al., 2022). The researchers have stressed the critical importance of information technology in medical tourism in terms of supporting visitors in both the pre-procedure and after-procedure phases of medical tourism. Unfortunately, the advantages of blockchain regarding pre- and after-procedure haven't yet been explored in depth. As a result, the researchers who conducted this research seek to envisage the potential of blockchain technology for healthcare tourism's before- and after-procedure stages. According to Ref, it is essential to concentrate on how blockchain may help both customers and vendors. This chapter intends to build on past studies and add to the current literature on the advantages of blockchains for medical tourism, especially for before- and after-procedure care. The chapter's remaining portion is laid out as follows (Esposito et al., 2018).

The second section is an examination of the literature on medical tourism, the role of computer technology in healthcare tourism, including blockchain. The final segment discusses the potential advantages of blockchains for health tourism. The fourth segment revisits the issues of health tourism in the aftermath of the epidemic and discusses the prospects for blockchain application. Lastly, the latter phase discusses the key findings as well as research directions.

11.2   LITERATURE REVIEW

Blockchain offers a number of built-in capabilities, such as distributed ledger, decentralized storing, identification, privacy, and data integrity, and that it has progressed past sensationalism to actual uses in sectors such as healthcare. Due to severe regulatory requirements, such as the Health Insurance Portability and Accountability Act of 1996, blockchain applications in the healthcare sector often require more stringent authentication, interoperability, and record-sharing requirements. Based on current blockchain technology, scientists in the academic and business world have begun to investigate possibilities for healthcare usage. Blockchain networks, spam detection, and biometric authentication are among these possibilities. Despite these advancements, there are still concerns since blockchain has its own set of weaknesses and difficulties to solve, such as processing motivations, extraction attacks, and strong authentication. Furthermore, as stated in this survey article, many healthcare applications have particular needs that are not covered by many of the blockchain trials being investigated. This study also discusses some prospective research avenues. In recent history, blockchain technology has demonstrated remarkable flexibility as various business industries found methods to incorporate its capabilities into their processes. While the financial services industry has received most of the attention thus far, initiatives in other business fields, such as healthcare, suggest that this is starting to change. This research focusses on a variety of starting points for blockchain technology in the healthcare industry. This chapter attempts to highlight various effects, ambitions, and potentials associated with this disruptive technology by using examples from public healthcare administration, user-oriented medical research, and medication counterfeiting in the pharmaceutical industry (Jamader et al., 2021).

Blockchain technology is one of the most important discoveries and creative advancements in the professional world today. Blockchain technology moves in the direction of continuous transformation. It is a network of blocks that protects information and preserves trust between people, despite how far apart they are. The recent growth in blockchain technology has compelled researchers and professionals to investigate innovative ways to utilize blockchain technology across a wide variety of fields. The exponential growth of blockchain technology has generated several new potential uses, particularly in smart healthcare. This research offers an in-depth examination of developing blockchain-based medical technology and applications. In this investigation, we draw attention to the outstanding methodological issues in this rapidly expanding subject and discuss them in depth. We also demonstrate how blockchain technology has the potential to transform the healthcare industry.

Obtaining, maintaining, combining, as well as exchanging health information securely are difficulties for patients and healthcare practitioners. Patients ought to be able to manage their health records from anywhere in the globe, maintain track of their medical history, provide access to data, and securely communicate this information with any healthcare practitioner. Concrete data access for consumers combined with a more comprehensive computation architecture might best prepare the medical industry to tackle public health concerns like COVID-19. Due to limits in confidentiality, as well as complete ecosystem interoperability, current technologies in use by the healthcare sector do not effectively fulfil these criteria. This research conducted a scientific assessment to determine the crucial responsibilities blockchain technology plays in addressing some of the most pressing as well as difficult challenges confronting the healthcare sector. This chapter analyses the obstacles and potential for integrating blockchain technology in healthcare, as well as a summary of wellness blockchain technologies, including significant organizations providing alternatives throughout various uses. By doing so, we enhance and supplement existing blockchain research in healthcare (Jamader et al., 2022).

Compatibility in health coverage has typically been centred on information sharing across commercial organizations, such as various medical centres. Conversely, there has been a new effort toward physician interconnection, wherein electronic health record interchange is between physicians as well as caregivers. Furthermore, customer connectivity introduces additional difficulties and needs in terms of security and privacy, technology, incentives, and governance that must be addressed for this sort of data exchange to function at scale. In this study, we examine how blockchain technology might help with this shift through five mechanisms:

(1) Electronic access restrictions,
(2) Known solution,
(3) Knowledge volatility,
(4) Diagnosis, and
(5) Information atomicity.

We next look at the challenges to distributed ledgers' physician compatibility, notably medical information activity volumes, confidentiality and security, treatment adherence, as well as compensation. We continue by emphasizing that, while patient-driven interconnection indeed offers innovative solutions in medicine, considering these limitations, it remains to be determined whether blockchain can aid the move from institution-centric to patient-centric data exchange (Jamader et al., 2023).

Blockchain technology eliminates the need for a centralized authority to authenticate data integrity but also provenance, as well as to facilitate interactions and the transfer of online services, while smartphones are equipped as well as pseudo anonymous interactions and arrangements among clients and providers. It contains crucial qualities such as irreversibility, decentralization, and openness, which have the potential to address critical healthcare concerns such as incomplete records at the point of service and limited access to patients' own healthcare information. Interoperability is required for an efficient and successful healthcare system because it allows software applications and technology platforms to connect safely and efficiently, share information, and utilize the data communicated between health organizations among application suppliers. Consequently, a lack of interoperability has resulted in segregated and fragmented data, delayed communications, and separate workflow tools in healthcare today. Blockchain has the potential to provide safe and pseudo-anonymous access to longitudinal, complete, and tamper-proof medical information held in fragmented systems. This chapter focusses on blockchain technology's application in healthcare and the implementation of blockchain technology in healthcare by

(1) Finding potential blockchain use applications in healthcare,
(2) Giving a case study that applies blockchain technology, and
(3) Assessing system design for using this technology in healthcare.

Blockchain technology has gotten a lot of attention recently, with increasing interest in a variety of applications spanning from data management, financial services, cybersecurity, IoT, and food science to the healthcare business and brain research. There has been a surge in interest in using blockchain technologies to enable safe and secure healthcare data management. Furthermore, blockchain is transforming traditional healthcare methods into more dependable ways of effective diagnosis and treatment through safe and secure data sharing. In the future, blockchain might be a technology that can aid with tailored, genuine, and secure healthcare by combining all of a patient's real-time clinical data and presenting it in an up-to-date secure healthcare setting. Blockchain might be a technology that can aid with tailored, genuine, and secure medicine by combining all pertinent patient information from a patient's health and presenting it in a secure healthcare setting. In this study, we use blockchain as a model to analyse both existing and recent innovations in healthcare. We also discuss blockchain uses, problems, and future prospects (Jamader, A. R., Chowdhary, S., & Shankar Jha, S., 2023).

Recently there has been an increase in requests for healthcare professionals to give people authority over their personal health information. Nonetheless, security concerns about how various healthcare providers transmit healthcare information have resulted in a failure to establish such systems. The capacity to securely transmit data is critical for patients to get innovative borderless integrated healthcare services. Blockchain technology, because of its decentralized nature, is an appropriate driver for the much-needed transition toward unified medical, giving new perspectives and tackling some of the most pressing difficulties in many healthcare fields (Jamader et al., 2023). Blockchain technology enables healthcare organizations to register as well as handle participant interactions over a system with no centralized power. In this chapter, we will look at the notion of blockchains as well as the barriers to their implementation in the healthcare industry. Furthermore, an evaluation of the most recent blockchain technology deployments in healthcare is undertaken. Eventually, a novel case study of a blockchain-based medical infrastructure is provided, with suggestions for future blockchain researchers and developers to resolve the shortcomings of current systems (Jamader, A. R., Das, P., & Acharya, B. R., 2019).

11.3 BLOCKCHAIN IMPLICATIONS IN HEALTHCARE

This enthusiasm and fervour has now spread to health informatics. To maximize the positive significance and relevance of blockchain in medical services, the Office of the National Coordinator for Health Information Technology held an ideation challenge in 2016 to collect white papers on blockchain's various uses in medical services. This task led to a variety of potential blockchain healthcare applications. While keeping whole patient records inside the blockchain might be considered a use application for medical services, numerous possible challenges to deployment have been noted, particularly violations of privacy, legal requirements, and technological barriers connected to information storage and transfer. As a result, most short-term approaches have concentrated on error detection, monitoring, and permission. Upper Portion, an information security startup located in the Netherlands, collaborated with the Estonian government to develop a blockchain-based system for validating patient IDs. Every citizen was given a smartcard that linked their electronic health data (EHR) to their blockchain-based identification. Any EHR change is issued a hash and recorded in the blockchain. This method assures that data in the EHR has an irreversible independent audit and that entries cannot be deliberately altered. The permanent, generally understood logs may also be used to archive the current state of material in conventional healthcare platforms (Jamader, A. R., 2022).

Any change to the medical databases, such as booking systems, is timestamped and algorithmically verified in a block. Given the recent emphasis on information honesty as a result of concerns about scheduling fraud at the Veterans Administration and the risk of data manipulation in implantable medical devices such as pacemakers, such a system could provide several potential advantages to ensure that any changes to the patient information are safe and open to scrutiny.

11.4   BLOCKCHAIN'S TECHNOLOGICAL RESTRICTIONS

When opposed to traditional data storage methods, there are several possible drawbacks, including potential concerns with the dissemination of personally identifiable healthcare data inside a shared database, expanding the technology, as well as the expenditure of deployment. To begin, while data within the blockchain can be deidentified and encrypted, distributed access to the complete data set carries the danger of corruption or provides a service. Secondly, because problems have already emerged in smaller blockchain-based applications, the speed and scalability of a totally distributed system would also need to be addressed. All transactions are expected to be maintained on each network interface in a blockchain implementation. As an example of a possible bottleneck, properly participating as a miner in the Bitcoin network necessitates downloading the whole Bitcoin ledger, which weighed in at 101 terabytes at the end of 2016. Furthermore, the Bitcoin network's maximum rate of transaction validation is 7 transactions per second, which may limit the throughput of large blockchain networks. The cost-effectiveness of such a platform that can contain substantially more data has yet to be demonstrated in operational scenarios. To establish if a return on investment for this technology can be attained, the total cost of hardware, implementation, and support must be calculated. Such obstacles suggest that, while blockchain has the potential to provide transparency and authenticity to data exchanges, swiftly switching current healthcare IT systems to blockchain-based technologies may be problematic (Mayakul, T., Kiattisin, S., & Prasad, R., 2018).

11.5 BLOCKCHAIN TECHNOLOGY EXPLANATION

Blockchain innovation has received a lot of interest because its initial use, the commodity Bitcoin, emerged as profitable and extremely famous on the worldwide market. Bitcoin, introduced in 2008 by the assumed mysterious Satoshi Nakamoto, was priced roughly $0.10 in 2009 and attained a high of $20,000 in December 2017, and is expected to achieve a new peak of around $44,000 in February 2021. In the article "Bitcoin: A Peer-to-Peer Digital Cash System," Phone Pay developed electronic cash, allowing internet transactions to be transmitted straight from one person to another without requiring the use of middlemen. Although most individuals associate blockchain with Bitcoin, blockchain technology is relevant to any electronic property transfer conducted over the internet. The success of Bitcoin has resulted in an exponential development of the crypto industry, with more and more alternative cryptocurrencies being introduced. Whereas blockchain innovation has been primarily used in the economic sector, it is discovering new possibilities in areas such as health, government, and tourism. Blockchain is a distributed database of chronological transactions that are stored in blocks linked to one another. In the blockchain network, all information is duplicated and distributed across multiple computers known as nodes. All parties must vote on whether the operations are reliable enough to be recorded on the blockchain. Furthermore, every endpoint is urged to verify every incoming transaction.

Because neither altering nor removing previously inputted entries requires considerable effort, the information can be considered safe. Every subsequent issue contains a hashing value produced from the transactional data as well as the preceding block's security number. If any changes are made, the hash value changes, making it easier to notice any changes in the blocks, therefore it is better to properly monitor the data. Figure 11.1 depicts a common blockchain work procedure.

Whenever a payment is requested, information is turned into a hashed block including data such as the date/time, sender, recipient, asset type, and amount, and saved as a candidate for publishing on the ledger. The block is then broadcast to a network of connected processors for validation. Once validated, the transaction is added to the chain of blocks, completing the transaction (Oberoi, S., Kansra, P., & Gupta, S. L., 2022).

Because of its unique properties, blockchain technology has the potential to transform several sectors. Among the many benefits of blockchain technology are:

1. **Elimination of Middlemen**: Blockchain's peer-to-peer structure minimizes the need for a centralized authority.
2. **Data Consistency**: Any data contributed to the blockchain cannot be edited or erased, resulting in an extremely high degree of security.
3. **Accountability**: To confirm their source and its route, all data may be tracked.
4. **Safety**: The chain's block cryptography combined with the provision of a unique identity for each blockchain user ensures a high level of security.

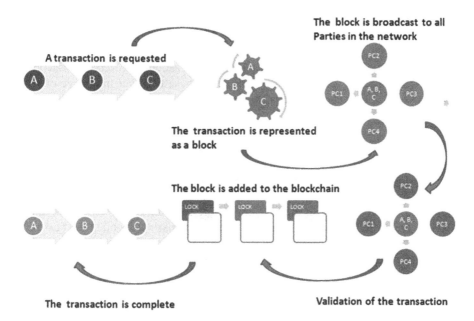

FIGURE 11.1   An example blockchain workflow procedure. Source: Authors.

5. **Quicker Processing**: Compared to the traditional banking method, which typically takes three days to execute a transaction, blockchain technology substantially cuts processing time to minutes or even seconds.
6. **Expense Savings**: The removal of middlemen allows for cost savings.
7. **Trust**: All blockchain users may trust one another and transact directly with one another.

Despite its benefits, blockchain technology does have certain drawbacks:

1. **Energy :Consumption**: Mining blocks and maintaining a real-time ledger consume massive amounts of energy.
2. **The Quantity of Energy**: In a given year, the energy consumption of solely Bitcoin miners exceeded the per capita energy consumption of a single nation.
3. **Expensive Operating Costs**: Small businesses consider Bitcoin to be unprofitable.
4. **The Regulatory Status Is Unknown**: The existing financial institutions' adoption and widespread usage of cryptocurrencies are hampered by a lack of regulation.

There are two kinds of blockchains: formal and informal blockchain. Everybody has access to the public blockchain, and everyone on the chain may participate in the general agreement mechanism.

Blockchain network is used by a restricted organization to preserve transactions that are only of relevance to subscribers of such blockchain. One more type of personal network is the consortium or consortium model, which is distinguished by pre-selected nodes that govern the communication protocol.

11.6   BLOCKCHAIN TECH'S ADVANTAGES IN HEALTH TOURISM

This segment discusses the advantages of blockchains for before- and after-procedure health tourism. Figure 11.2 summarizes the advantages of blockchain technology for medical tourism. The initial phase of healthcare services is the before-procedure phase, which usually comprises a medical tourist preparing to obtain medical services. Looking for more information, locating a health tourism coordinator, arranging accommodation as well as planning processes, pre-verification, and medical intervention are all crucial phases in this step. Certain parts of the medical tourism pre-procedure can be facilitated by blockchain-enabled disintermediation and blockchain-based cryptocurrencies and transactions.

The early stage of health tourism is the before-procedure phase, which comprises a healthcare traveller preparing to obtain healthcare. Collecting data, locating a health tourism mediator, arranging logistics and a plan of action, pre-medical

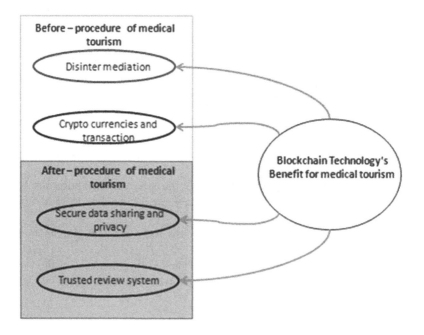

FIGURE 11.2   The advantages of blockchains for health tourism. Source: Authors.

check-up, and medical intervention are all crucial phases in this segment. Certain parts of the medical tourism pre-procedure can be facilitated by blockchain-enabled disintermediation and blockchain-based cryptocurrencies and transactions. With the expansion of the medical tourism business, several travel firms have transformed into medical tourism enablers, acting as mediators between healthcare providers and medical tourists. Because healthcare systems always can determine the suitability and quality of a medical tourism location, they must seek out and depend on private healthcare intermediaries to address their concerns and plan their schedules. However, international practitioners must contract medical tourism packages at costs higher than the real cost of medical service providers and are limited to the available selections owing to the medical tourism facilitator's relationship with specific medical providers.

A further concern is whether medical tourism facilitators are legally liable for any faults in medical treatment. Due to its capacity to develop trust, provide more secure information transmission, lower expenses, and promote transparency, blockchain technology has the potential to dramatically reduce or eliminate the influence of medical tourism facilitators. As a result, blockchain technology can allow international practitioners to communicate directly with medical service providers. In terms of healthcare providers, the ledger system offers equal opportunities for the largest hospitals as well as smaller and less popular medical service providers to target potential patients. Because blockchain protects the source, authenticity, and openness of data, medical tourists may examine the credentials and certificates of medical service providers and be confident that the operational fees are uniform for everyone (Paris, C. M., 2022).

Ethereum-based payments as well as interactions should help with the procedures that need to be resolved in the area of health tourism. Bitcoins enable simple, quick, relatively secure monetary exchanges without the requirement for a reliable third party. As a result, both foreign patients and medical service providers might benefit from safe transactions, ease of use, high productivity, shorter waiting lists, and safe storage of patient data. Furthermore, Bitcoin-associated payments provide efficiency in international transfers with no processing costs or movement charges to medical tourists, giving a competitive edge to doctors.

11.7 ADVANTAGES OF BLOCKCHAIN IN HEALTH TOURISM

The second process of health tourism is after-procedure maintenance, which includes post-operative treatment and follow-up care for medical tourists. This section encompasses the following phases: morbidity surveillance, exercise rehabilitation as well as advancement checks; take concern, drug admonition as well as post-treatment free time; repayments approval as well as fly home, but also neighbourhood general practitioner support. Blockchain can be used to establish a safe registry for healthcare visitors' medical files, as well as a reliable feedback process.

The effective and secure exchange of medical records is critical for medical tourists' post-operative care and concerns. Medical tourists, foreign medical service providers, local primary healthcare practitioners, as well as other stakeholders, can benefit from the effective and rapid sharing of health data enabled by blockchain technology. Blockchains can be used to create an ideal cloud for documenting, storing, and managing the health data and medications of medical tourists. The blockchain system's interoperability allows healthcare visitors to connect their medical status and registers via a decentralized system. Furthermore, medical professionals can share data with a general practitioner to keep the healthcare continuum intact. Blockchain technology ensures the integrity of medical information due to its immutability and traceability. Neither the medical service provider nor the medical tourists have the ability to change or delete the implanted data. Furthermore, in order to maintain confidentiality, blockchain technology allows medical tourists to retain control over their hospital data. As a result, foreign patients choose which data to share, with whom, and how. Because visitors are fully aware of the flow and use of their personal information, their privacy concerns are likely to diminish (Pilkington, M., 2020).

When choosing a medical tourism destination or medical service provider, prospective medical tourists frequently search for evaluations. In the meantime, foreign patients who have previously received medical services are more likely to leave feedback about their experiences. Blockchain technology, by enabling trusted review systems, has the ability to guarantee the trustworthiness and authenticity of the evaluations. Including several verifiable proof procedures integrated into the review systems, blockchain technology is capable of providing a unique private key for each identity. This could mean fewer instances of review manipulation or duplication. A few foreign patients avoid leaving evaluations because they do not want to be identified. This problem, nevertheless, can be addressed by blockchain-based systems that can keep critics pseudonymous. Furthermore, smart contracts research surveys could indeed recompense medical tourists with coins or tokens for leaving reviews and sharing their experiences.

11.8 BOUNDARIES OF BLOCKCHAIN BRING INTO PLAY IN MEDICAL TOURISM

When applied in the private healthcare sector, blockchain technology has limitations that must be considered. One of the most fundamental limits is related to data storage and management. The data integrity feature of blockchain results in irreversibility, which means that once information has been entered into the Bitcoin protocol, it can't be removed or changed. Nevertheless, because healthcare data is protected by privacy laws, it must be deleted if medical tourists request it. As a result, anyone considering using blockchain to contain health records should reconsider. Furthermore, while blockchain technology is ideal for recording medical tourists' data, it is not suitable for storing large amounts of information or supersonic speed information because of the enormous duplication among many highly interconnected processing elements trying to hold a comprehensive copy of the entire data. To circumvent this restriction, even hashes and other such concepts can be stored on the blockchain, while primary data is stored off-chain. Furthermore, if cyber events happen, they will be extremely difficult to reverse.

Because while the ledger maintains safety and confidentiality, the safety of investments is dependent on safeguarding the private key, which is a form of electronic identity. If a secret key is stolen or misplaced, no foreign entity can recover it, resulting in the loss of both information and assets (Rana, R. L., Adamashvili, N., & Tricase, C., 2022).

To safeguard a secret key, each chain's authentication mechanism must be linked to some other Bitcoin fingerprint-collected data. A further barrier to blockchain adoption is the lack of standardization of blockchain frameworks. Due to the problems in integrating different architectures, this can impede the establishment of business connections between healthcare providers by implementing blockchain architectures. Furthermore, regulatory constraints in practical implementations inhibit smart contracts from being implemented in a number of countries. Finally, there is constantly the issue of sustainability with reference to energy wastage. Public blockchains would then use much more power than centralized, non-replicated data sources. Experts and stakeholders, on the other hand, are working to overcome the constraints of distributed ledger technology. Recent legislative measures, organizational partnering, as well as the advancement of more effective blockchain frameworks, are just a few instances of how blockchain adoption can be aided.

11.9 PROSPECTS FOR BLOCKCHAIN USAGES DUE TO THE COVID-19 DISEASE OUTBREAK

The COVID-19 global epidemic has had a massive impact on human inhabitants as well as nearly all enterprises and markets. Health tourism has been affected worldwide in various ways. Numerous nations have limited mobility and imposed long-term lockdowns, curfews, and other forms of social segregation. Many airlines, especially those to popular medical tourism destinations, have been cancelled, along with local and intercity connectivity, and several tourist industry enterprises. COVID-19 has had a significant impact on the healthcare system in medical tourism destinations. Many hospitals have had to treat COVID-19 cases and develop COVID-19 recovery units instead of providing hospital rooms or treatment units for international medical tourists. Most hospitals could only perform safety protocols; non-essential procedures were delayed or cancelled. Moreover, a healthcare provider's COVID-19 safety measures are critical for maintaining the outstanding quality and innovativeness regarded by world healthcare visitors. Despite the difficulties of a disease outbreak for the medical tourism market, several opportunities for blockchain use may exist. For starters, blockchain technology is more likely to be used as a secure database for data storage and dissemination from patients' health records among multiple patients and medical professionals. In terms of encouraging telemedicine, the pandemic forced the digitization of the medical tourism industry. Physicians from all over the world can discuss and treat both national and international clients using content and teleconferencing techniques. There are times when the views of many physicians must be discussed, or a practitioner must be substituted. In these cases, the physician can send the necessary information to the doctors in a safe and timely manner.

Primarily, blockchains can be employed to create as well as validate SSL signature verifying medical tourism service providers' high-quality certification. In the event of a disease outbreak, foreign patients are more likely to prefer national medical tourism providers that have been accredited to ensure compliance with World Health Organization guidelines and public healthcare applicable regulations. Blockchain-based systems will confirm credential financial institutions as well as beneficiaries, focus on ensuring credential truthfulness, and expose forged accreditations. Furthermore, blockchains may be used to create virtual credentials containing PCR findings and other wellness documentation. PCR exam credentials are among the minimum standards which a traveller must meet when leaving and attempting to enter a nation. Blockchain could indeed remove credential forgery and exploitation, ensuring their truthfulness. Furthermore, the visitors will have total power over the documents and will gain entry to officials who require them (Srivastava et al., 2021).

Eventually, a distributed ledger has the potential to issue healthcare vacationers' visas. Visas are frequently required for medical tourists visiting a healthcare travel destination. Visa processing takes time, and in an emergency, medical tourists do not have much time to wait for a residence permit. Blockchain has the ability to significantly reduce wait times by effectively managing all transport official documents, such as residence permit granting.

11.10 CONCLUSION

The primary goal of this chapter was to expand on past studies as well as enrich the existing publications on the advantages of blockchain technology in healthcare tourism. Earlier studies investigated the impact of data innovations in medical tourism before and after procedures; nevertheless, they did not concentrate on the beneficial role of blockchains. To add to the body of knowledge, this chapter primarily focused on the potential strategic role of blockchain technology and conceptualized the benefits of blockchain technology for two major stages of health tourism: during and after the process.

As a result, the chapter contends that medical tourists and healthcare providers alike may benefit from the use of blockchain technology in terms of an easier way to find a healthcare provider, a faster and more secure payment method, guaranteed data security and privacy, and trusted and highly recommended. Furthermore, the chapter extends the current literature on blockchain in tourism research by suggesting the future consequences of these modern features for the medical travel industry. Given the COVID-19 disease outbreak, the chapter is both timely and pertinent. The pandemic had an impact on the entire tourism industry, including medical tourism. To address this issue, the chapter summarized several challenges that the medical tourism industry is facing as a result of the pandemic and outlined some opportunities for blockchain technology use in the current situation, such as reducing waiting times for all travel-related government documents, including residence permit granting, as well as ensuring the authenticity of these records.

Whereas the chapter provides many perspectives on the advantages and potential of blockchain technology for health tourism, it does have some constraints.

The extent is restricted because the chapter just concentrates on many phases of medical travel before and after procedure. Furthermore, a more in-depth qualitative approach, as well as empirical research, is required to improve our knowledge of the potential of blockchains in the medical tourism business. It will be critical to undertake evaluation research to discover additional potential areas of healthcare services that might be impacted by the use of blockchain technology. Furthermore, investigators must confront the difficulties of innovation as well as the threat of using blockchain technology in the medical tourism industry.

REFERENCES

Barkan, M., & Tapliashvili, N. (2018). Cryptocurrency Use in Medical Tourism. *Scientific and Practical Cyber Security Journal (SPCSJ)*, 2(4), 104–110.

Das, P., Jamader, A. R., Acharya, B. R., & Das, H. (2019, May). HMF based QoS aware recommended resource allocation system in mobile edge computing for IoT. In *2019 International Conference on Intelligent Computing and Control Systems (ICCS)* (pp. 444–449). IEEE.

Das, P., Martin Sagayam, K., Rahaman Jamader, A., & Acharya, B. (2022). Remote Sensing in Public Health Environment: A Review. *Internet of Things Based Smart Healthcare*, 379–397. https://doi.org/10.1007/978-981-19-1408-9_17

Esposito, C., De Santis, A., Tortora, G., Chang, H., & Choo, K. K. R. (2018). Blockchain: A Panacea for Healthcare Cloud-Based Data Security and Privacy? *IEEE Cloud Computing*, 5(1), 31–37.

Jamader, A. R. (2022). A Brief Report of the Upcoming & Present Economic Impact To Hospitality Industry in COVID19 Situations. *Journal of Pharmaceutical Negative Results*, 2289–2302. https://doi.org/10.47750/pnr.2023.14.s02.202

Jamader, A. R., Das, P., Acharya, B., & Hu, Y. C. (2021). Overview of Security and Protection Techniques for Microgrids. In *Microgrids* (pp. 231–253). CRC Press.

Jamader, A. R., Das, P., & Acharya, B. (2022). An Analysis of Consumers Acceptance towards Usage of Digital Payment System, Fintech and CBDC. *Fintech and CBDC (January 1, 2022)*.

Jamader, A. R., Chowdhary, S., Jha, S. S., & Roy, B. (2023). Application of Economic Models to Green Circumstance for Management of Littoral Area: A Sustainable Tourism Arrangement. *SMART Journal of Business Management Studies*, 19(1), 70–84.

Jamader, A. R., Chowdhary, S., & Shankar Jha, S. (2023). A Road Map for Two Decades of Sustainable Tourism Development Framework. In *Resilient and Sustainable Destinations After Disaster: Challenges and Strategies* (pp. 9–18). Emerald Publishing Limited.

Jamader, A. R., Das, P., & Acharya, B. R. (2019, May). BcIoT: Blockchain based DDoS Prevention Architecture for IoT. In *2019 International Conference on Intelligent Computing and Control Systems (ICCS)* (pp. 377–382). IEEE.

Jamader, A. R., Immanuel, J. S., Ebenezer, V., Rakhi, R. A., Sagayam, K. M., & Das, P. (2023). Virtual Education, Training and Internships in Hospitality and Tourism During Covid-19 Situation. *Journal of Pharmaceutical Negative Results*, 286–290.

Mayakul, T., Kiattisin, S., & Prasad, R. (2018). A Sustainable Medical Tourism Framework Based on the Enterprise Architecture Design: The Case in Thailand. *Journal of Green Engineering*, 8(3), 359–388. https://doi.org/10.1080/10963758.2021.1907198

Nayak, D. K., Mishra, P., Das, P., Jamader, A. R., & Acharya, B. (2022). Application of Deep Learning in Biomedical Informatics and Healthcare. In *Smart Healthcare Analytics: State of the Art* (pp. 113–132). Springer.

Oberoi, S., Kansra, P., & Gupta, S. L. (2022). A Systematic Review on Determinants Inciting Sustainable E-Medical Tourism. *International Journal of Reliable and Quality E-Healthcare (IJRQEH)*, *11*(2), 1–13.

Paris, C. M. (2022). Leveraging Blockchain in Medical Tourism Value Chain. In *Information and Communication Technologies in Tourism 2022: Proceedings of the ENTER 2022 ETourism Conference, January 11–14, 2022* (p. 78). Springer Nature.

Pilkington, M. (2020). The Relation between Tokens and Blockchain Networks: The Case of Medical Tourism in the Republic of Moldova. *The Journal of The British Blockchain Association*, 18156. https://doi.org/10.31585/jbba-4-1-(2)2021

Rana, R. L., Adamashvili, N., & Tricase, C. (2022). The Impact of Blockchain Technology Adoption on Tourism Industry: A Systematic Literature Review. *Sustainability*, *14*(12), 7383.

Sagayam, K. M., Das, P., Jamader, A. R., Acharya, B. R., Bonyah, E., & Elngar, A. A. (2022). DeepCOVIDNet: COVID-19 Detec

Srivastava, A., Jain, P., Hazela, B., Asthana, P., & Rizvi, S. W. A. (2021). Application of Fog Computing, Internet of Things, and Blockchain Technology in Healthcare Industry. In *Fog Computing for Healthcare 4.0 Environments* (pp. 563–591). Springer, Cham.

Index

Printed in the United States
by Baker & Taylor Publisher Services